Studies of Brain Function, Vol. 8

Studies of Brain Function

Volumes already published in the series:

Juhani Hyvärinen

The Parietal Cortex
of Monkey and Man

With 85 Figures

Springer-Verlag
Berlin Heidelberg New York 1982

Dr. JUHANI HYVÄRINEN, Department of Physiology
University of Helsinki, Siltavuorenpenger 20 J
00170 Helsinki 17, Finland

ISBN 3-540-11652-4 Springer-Verlag Berlin Heidelberg New York
ISBN 0-387-11652-4 Springer-Verlag New York Heidelberg Berlin

Library of Congress Cataloging in Publication Data. Hyvärinen, Juhani. The
parietal cortex of monkey and man. (Studies of brain function; v. 8) Bibli-
ography: p. Includes index. 1. Parietal lobes. 2. Monkeys – Physiology.
I. Title. II. Series. [DNLM: 1. Parietal lobe – Physiology. Wl ST937KF f. 8/
WL 307 H999p] QP382.P3H96   612′.825   82-10256

Offsetprinting and binding: Konrad Triltsch. Graphischer Betrieb, Würzburg
2131/3130-543210

Dedicated to Vernon B. Mountcastle

Preface

An invitation from the Editors to contribute to *'Studies of Brain Functions'* with a monograph on the parietal lobe offers me an opportunity to present in a concentrated form my studies on this part of the brain from a period of somewhat over a decade. The parietal lobe, notably its posterior part, is a very complex neural system whose functions I have been able to study only superficially and without extensive coverage of all its parts. Therefore I did not want to limit myself entirely to my own work but found the task of writing more interesting by including sections reviewing relevant literature. Thus Chapter III dealing with the primary somatosensory cortex and Chapters IX, X, and XI concerning area 7 describe work done in my laboratory. Chapter VIII describes microelectrode work on area 7 and covers both the work of my group and that of others working on this area. Chapters II and IV to VII are based on closely related anatomical, physiological and clinical studies performed by others, and Chapter XII is a personal attempt at a synthesis of the functions of the parietal lobe. Thus this monograph is neither a strict review of all important works on the parietal lobe nor is it limited only to my own studies and those of my collaborators. Instead it attempts to be a balanced exposition of both aspects promoting, hopefully, a synthetic view of the primate parietal lobe.

The writing of this monograph, as well as many of the studies described here, have been financially aided by the National Research Council for Medical Sciences, Finland, and the Sigrid Juselius Foundation, Helsinki, to whom I express my gratitude. My collaborators in the studies described in this book Drs. Antti Poranen, Lea Leinonen, Anssi Sovijärvi, Synnöve Carlson, Lea Hyvärinen, Mr. Ilkka Linnankoski and Mrs. Katriina Laurén also deserve my sincere thanks for their contribution, without which this work would not have been possible. Many publishers have given permission to quote previous work and reprint figures, which is gratefully acknowledged. My thanks are also due to Mrs. Seija Turunen for help in locating the literature cited and Mrs. Leena Manner for assistance in the computerized typing of the manuscript.

Helsinki, August 1982   Juhani Hyvärinen

Contents

I. Introduction

MacDonald Critchley in the introduction of his admirable book *The Parietal Lobes* (1953) stated that there is something essentially artificial in extensive elaboration limited to one sector of the brain such as the parietal lobe. For this reason he was tempted to predict that his book might be the last one on this topic and that subsequent writers would approach the problem of cerebral fuction from a wider angle. I am not unhappy to belong to those followers of his who disprove this prediction and write another book on the parietal lobe. On one hand the present knowledge of the functional mechanisms of the parietal lobe alone constitutes a vast field of scientific inquiry that well deserves such treatment. On the other hand the knowledge gained from studies of this part of the brain have great significance for neuroscience as a whole. It is true that the parietal lobes are not working in isolation from other parts of the CNS and, in spite of certain specialization, they should not be viewed as seperate organs. However, the problems dealt with in studies of the parietal lobes illustrate general problems of brain research, particularly the complexity of functional mechanisms in the alert active brain. Although books on the parietal lobe have been slow to appear after Critchley's veritable accomplishment, recent basic and clinical studies have revived the interest and resulted in achievements that have given guidelines for future research. Incidentally, the studies done 100 years ago by such eminent pioneers as Ferrier (1876), Munk (1881) and others had great impact in the development of neuroscience. It appears likely that an impact extending far into the future will also be exercised by the studies of neuronal function in association cortex in behaving animals. The presentday findings may appear meagre when viewed decades from now, but the problems that are posed will undoubtedly interest neuroscientistis for a long time to come. Thus I do not share the doubtful prediction of Critchley about further books on parietal lobes, although books on other parts of the brain and more comprehensive systems will probably intervene in the series.

The studies on the parietal lobe fall into two main categories, those concerning the anterior parietal lobe or the *primary somatosensory cortex (SI),* and those dealing with the *posterior parietal* lobe or *associa-*

tion cortex. Studies on the *second somatosensory cortex (SII)* in the parietal operculum can best be grouped together with those of SI, whereas *area 5 (occasionally called SIII)* clearly belongs to the posterior parietal cortex. From the historic point of view the understanding of the somatosensory cortical regions has been coupled with the advancement of basic science, whereas for the development of ideas about the function of posterior parietal cortex clinical studies on patients were for a long time the most important source of inquiry. On the other hand, electric stimulation studies in man have contributed greatly to the understanding of somatic sensory cortical representation (Penfield and Rasmussen 1950) but revealed very little of posterior parietal function. Studies of evoked potentials (Adrian 1941, Woolsey et al. 1942) have delineated the somatosensory representations in animals, whereas the electric stimulation method proved more appropriate in humans for this purpose. Later microelectrode recording of cellular action potentials, particularly by Mountcastle and his associates, have advanced our understanding of the funtion of both the anterior (Powell and Mountcastle 1959b) and posterior (Mountcastle et al. 1975) parts of the parietal lobe.

II. Anatomy and Evolution of the Parietal Lobe in Monkeys and Man

A. Anatomy

From the evolutionary point of view the anterior and posterior parietal lobes differ from each other. The main structure of the anterior parietal lobe is the *primary somatosensory cortex (SI)* which consists in both monkeys and man of *Brodmann's cytoarchitectural areas 3, 1 and 2.* The basic topological representation of the body also appears similar in monkeys and man. The *posterior parietal lobe,* however, has developed further in man than in monkeys. It consists of a *superior and inferior lobule* in both species. The inferior parietal lobule in particular has enlarged in humans, consisting of *Brodmann's* (1907) *areas 40 and 39* which have no numerical counterpart in the monkey's brain. However, the inferior parietal lobules of monkey and human appear homologous because in von Economo's (1929) terminology for human brain and the corresponding terminology of von Bonin and Bailey (1947) for the monkey's brain the inferior parietal lobule consists of *areas PG and PF* in both species. The cytoarchitectural maps presented in Figs. 1 and 2 illustrate the views of different cytoarchitectonists. Figure 1 illustrates the human brain. In Brodmann's terminology the anterior parietal lobe consists of areas 3, 1 and 2, whereas the posterior part consists of areas 40 and 39. The Vogts (1926) agreed with Brodmann on the anterior part of the parietal lobe distinguishing areas 3a, 3b, and a region called 1+2. The corresponding terminology of von Economo is PA (for 3a), PB (for 3b), PC (for 1), and PD (for 2) (P stands for parietal). Figure 2 presents the cytoarchitectural areas of the monkey's brain as viewed by different authors. Brodmann (1905) and the Vogts (1919) made their maps of the *Cercopithecus* brain, whereas Bonin and Bailey (1947) transcribed von Economo's system of letters to the brain of the rhesus monkey *(Macaca mulatta).* They agreed with previous authors except that they could not distinguish a separate area 2 or PD.

Agreement on the posterior parietal cortex is less evident, however. In the human brain (Fig. 1) Brodmann (1907) subdivided the superior parietal lobule into areas 5 and 7 and the inferior parietal lobule to areas 39 and 40. The Vogts (1926), on the other hand, considered the entire

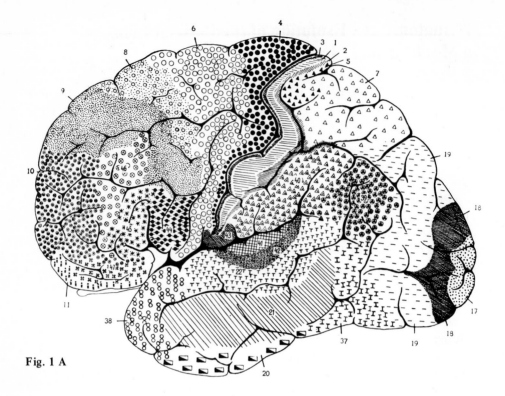

Fig. 1 A

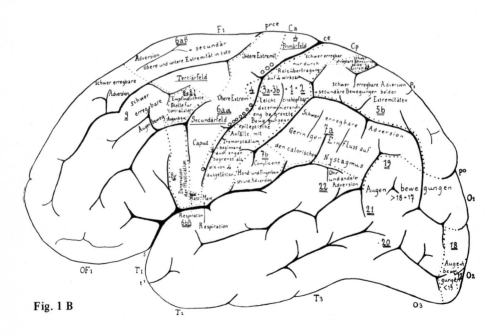

Fig. 1 B

superior parietal lobule area 5, dividing it into subareas 5a and 5b. In their view the human inferior parietal lobule consisted of areas 7a and 7b. Von Economo's (1929) corresponding areas were PE in the superior parietal lobule and PG and PF in the inferior parietal lobule.

In the monkey's brain (Fig. 2) the superior parietal lobule consists of area 5 (Brodmann 1905) subdivided into 5a and 5b by the Vogts (1919). Von Bonin and Bailey's (1947) corresponding term is PE with an anterior subdivision PEm. The inferior parietal lobule of the monkey consists of area 7 (Brodmann 1905) subdivided into areas 7a and 7b by the Vogts (1919) and called PG and PF by von Bonin and Bailey (1947).

On the basis of the maps of Figs. 1 and 2 it is obvious that von Bonin and Bailey (1947) considered all areas of the parietal lobe of the monkey homologous with the corresponding areas in the human brain. For each area in von Economo's nomenclature for man there is a corresponding area in the monkey's brain. The Vogts (1919, 1926) had also followed this principle, calling the inferior parietal lobule areas 7a and 7b both in the human and the simian brain. McCulloch (1944) appreciated the homology

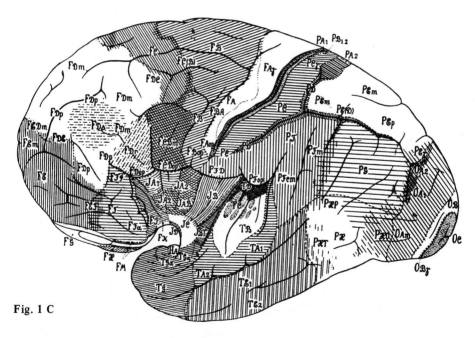

Fig. 1 C

Fig. 1 A-C. The cytoarchitectural areas of the lateral side of the human brain according to three different authors. **A** Brodmann's (1907) map. **B** the Vogts' (1926) map. The Vogts marked on the figure results of electrical stimulation that they expected could be elicited from human cortex on the basis of their stimulation results from monkeys. **C** von Economo's (1929) map

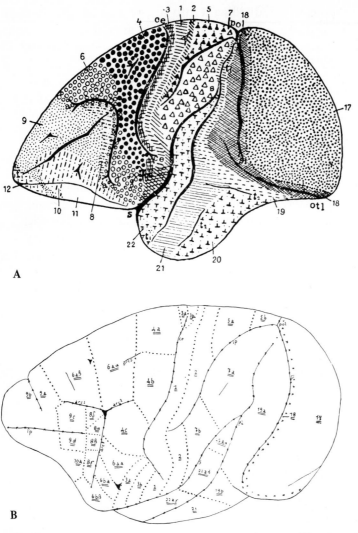

Fig. 2 A-C. The cytoarchitectural areas of the lateral side of the monkey brain according to **A** Brodmann (1905, *Cercopithecus*), **B** the Vogts (1919, *Cercopithecus*), and **C** von Bonin and Bailey (1947, *Macaca mulatta*). In this figure the boundaries between areas are indicated with *lines*, but Bonin and Bailey emphasized their gradual nature

so much that, using Brodmann's human terminology for the monkey, he subdivided the monkey's superior parietal lobule into areas 5 and 7 and the monkey's inferior parietal lobule into areas 39 (7a) and 40 (7b). Homologues of the human areas 39 and 40 are thus lacking only in Brodmann's original maps for the *Cercopithecus* brain (1905). As pointed out

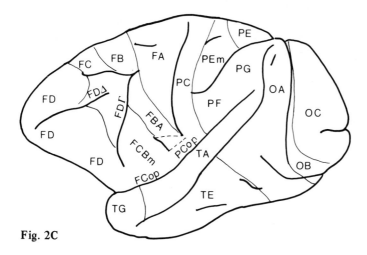

Fig. 2C

by von Bonin and Bailey (1947), Brodmann's (1907) original description of the cytoarchitecture of the human brain was scanty, for which reason they adopted the lettering system of von Economo.

In recent years Selzer and Pandya (1978, 1980) have described two new subdivisions of the monkey's posterior parietal cortex, *areas POa* in the intraparietal sulcus and *PGa* in the superior temporal sulcus (Fig.3). Moreover, Pandya and Selzer (1982) subdivided the posterior partietal lobe into several smaller cytoarchitectonic subdivisions which may turn out to be functionally specialized.

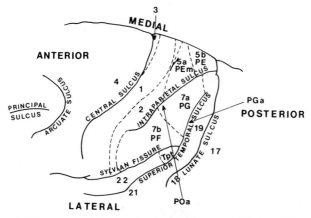

Fig. 3. The cytoarchitectural areas of the parietal lobe of the monkey with the designations used in this monograph. The areas buried in the sulci are indicated with *arrows*. Based on von Bonin and Bailey (1947), the Vogts' (1919), Powell and Mountcastle (1959a), Pandya and Sanides (1973) and Selzer and Pandya (1978, 1980), and modified from Hyvärinen (1982)

B. Evolution

The monkey (Fig. 4), being a primate, has two hands that it uses ag-
ilely for locomotion in trees and for picking up food and eating it. However, monkeys do not generally use their hands for work with tools. *Chimpanzees*, too, use their hands and arms for locomotion but they are also
known to use primitive tools. However, their hands are quite clumsy,
as can be seen in Fig. 5, where a chimpanzee is using a straw for picking
up termites from their nest.

The hand developed to the dexterous working organ only after it was
freed from the task of assisting locomotion. This phase of development
took place when man's early ancestors appeared on the earth. Their
essential advantage in comparison with the chimpanzee and other primates
was their ability to walk on two feet, which freed the hands for carrying
food for the family at the home base, manual use of tools and linguistic
communication. These developments were associated with an enlargement
of the parietal lobe, particularly its posterior and inferior parts. Holloway
(1968, 1974, 1976, 1978) has studied fossils of early hominids, particularly endocranial casts made from plastic materials poured into the skulls.

Fig. 4. The stumptail macaque used in our studies is seen here reaching with the hand
for a food reward

Fig. 5. A chimpanzee using a straw as a tool for picking up termites from their nest. (van Lawick-Goodall 1971)

A   B   C   D   E

Fig. 6 A-E. An artist's conception of evolution of man from earlier primates.
A *Dryopithecus africanus* (lived about 15 million years ago). **B** *Ramapithecus* (10 million years ago). **C** *Australopithecus africanus* (3 to 5 million years ago). **D** *Homo erectus* (0.5 million years ago). **E** *Homo sapiens* of the Neandertal type (50000 years ago)

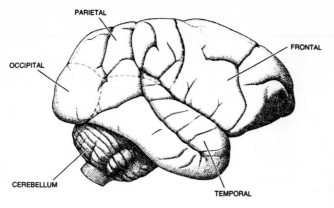

Fig. 7 A

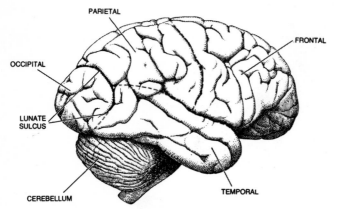

Fig. 7 B

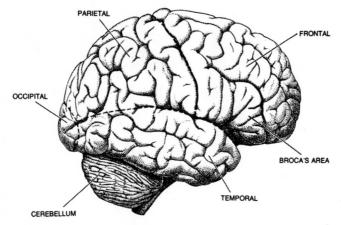

Fig. 7 C

Fig. 7 A-F. The evolution of the human brain according to Holloway (1974). On the left from top downwards the brains of **A** the cebus monkey, **B** the chimpanzee and **C** modern man. On the right endocranial casts of the skulls of **D** the chimpanzee, **E** *Australopithecus africanus* and **F** *Australopithecus robustus*

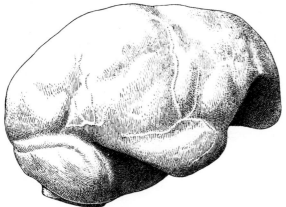

Fig. 7 D

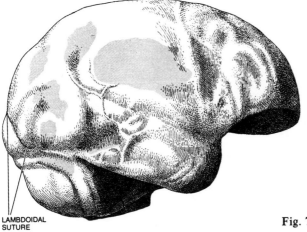

LAMBDOIDAL
SUTURE

Fig. 7 E

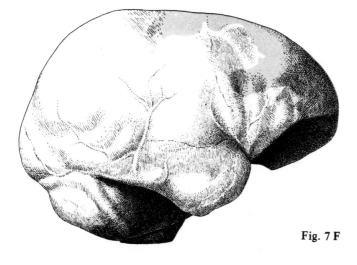

Fig. 7 F

One of the targets of this research was the *Australopithecus*, an early hominid that lived 3 to 5 million years ago (Fig. 6). Australopithecus was about the same size as the chimpanzee and had roughly the same brain size. However, he already walked on two feet and used tools. This phase of development was associated with an enlargement of the parietal lobe as is indicated by the more posterior location of the lambdoidal suture marking the border of the occipital and parietal lobes (Fig. 7). Thus the Australopithecus had a larger parietal lobe than the chimpanzee who, on the other hand, has a larger occipital lobe. Thus the Australopithecus had the prerequisites in the parietal lobe for more agile use of tools. However, the chimpanzee may be more efficent in visual performance than the Australopithecus was.

III. Functional Properties of Neurones in the Primary Somatosensory Cortex

The primary somatosensory cortex (SI) receives its main input from the *thalamic ventrobasal complex.* In the monkey the projection to area 3b is heavy and made up of singularly coarse fibres, whereas the fibres to areas 1 and 2 are much fewer in number and finer in calibre. It is possible that the fibres to areas 1 and 2 are branches of the coarser fibres passing to area 3b (Jones and Powell 1970b). The most anterior part of SI, area 3a, also receives a projection from the ventrobasal complex from the subnucleus ventralis posterolateralis, pars oralis (VPLo) (see Jones and Powell 1973).

Figure 8 shows a sagittal section across the central sulcus and indicates the antero-posterior location of the cytoarchitectural areas of the post-central gyrus. Area 3a is the adjoining area between the sensory and motor cortices at the bottom of the central sulcus. It differs from the more posterior parts of SI in that it receives a strong afferent input from muscles (Phillips et al. 1971), whereas the muscles have little or no representation in areas 3b and 1.

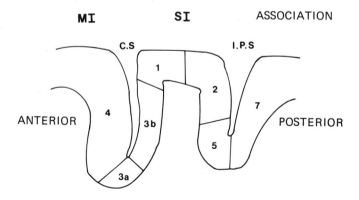

Fig. 8. A schematic drawing of a section across the central sulcus at the level of the hand representation in the monkey. Area *4* is the motor cortex, areas *3a, 3b, 1,* and *2* form the first somatosensory cortex; areas *5* and *7* belong to the posterior parietal association cortex. *C.S.* central sulcus; *I.P.S.* intraparietal sulcus. (Hyvärinen and Poranen 1978b)

It has previously been shown by Powell and Mountcastle (1959b) that the anterior part of SI is strongly connected with *cutaneous receptors*, whereas the posterior part is related to *receptors in joints and deep tissues*. Moreover, the projection from the periphery to the cortex follows a *dermatomal arrangement* with the rostral dermatomes represented laterally and the caudal ones medially (Werner and Whitsel 1968). The *columnar organization* of SI was first demonstrated by Mountcastle 1959b). According to the original concept of columnar organization, the cortex consists of a mosaic of narrow vertical columns; they extend perpendicularly across the cortex, each column containing functionally similar neurones. Columnar organization of the visual cortex was later studied extensively by Hubel and Wiesel and their collaborators, who have shown that in the visual cortex the "columns" are actually elongated "slabs" (Hubel and Wiesel 1977).

As described previously, most neurones in areas 3b´and 1 respond to cutaneous stimuli. Some are related to rapidly adapting cutaneous receptors, others to slowly adapting cutaneous receptors, and some respond to excitation of the subcutaneous Pacinian corpuscles. Neighbouring neurones encountered during an electrode penetration in SI

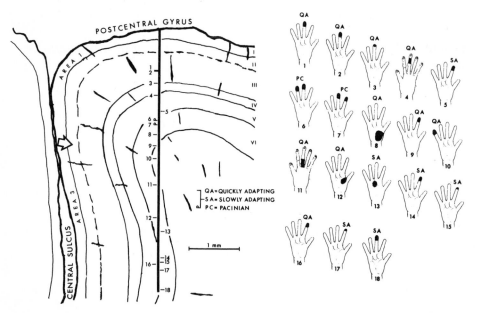

Fig. 9. An example of an electrode penetration across areas 1 and 3 b in the hand area of the monkey. The reconstructed electrode track is shown as a *black vertical line* along which are marked the 18 neurones recorded during this penetration. The figurines on the right indicate the receptive fields (*black*) and the submodalities of the neurones encountered. (Hyvärinen et al. 1969)

usually have similar or overlapping peripheral receptive fields; if the penetration is perpendicular to the cortex, they are related to similar peripheral receptors. An example of this arrangement is seen in Fig. 9, which shows a reconstruction of an electrode track across the cortical hand representation of the rhesus monkey. This penetration went diagonally across areas 1 and 3b, encountering within short distances neurones with similar receptive field properties before moving among other neurones with different properties. When such a penetration traverses the cortex in the direction of the radiation of the incoming fibres, the functional properties of the neurones encountered may be similar during the whole penetration, and the receptive fields may be closely overlapping (Powell and Mountcastle 1959b).

One of the main tasks of the primary somatosensory cortex is probably to convey messages about the location and timing of somatic stimuli. For this purpose responsiveness to touch and other simple somatic stimuli is enough. Thus the majority of the neurones encountered in SI show simple functional properties and respond whenever a stimulus excites the receptive field. However, in natural life both animals and man use the somesthetic sense in connection with active exploratory movements. The purpose of such exploration is often to identify and locate an object by manipulation or to get hold of it. In such tasks the stereognostic capacity of the somesthetic sense to synthetize three-dimensional representations of objects is put into practise. The visual system contains a large number of neurones that are sensitive to lines and edges and specific to their orientation (Hubel and Wiesel, 1962, 1968). Such functional features may be useful for the visual perception of form and patterns, and I was interested in studying whether in the primary somatosensory cortex there were comparable neuronal mechanisms to be found.

A. Comments About Methods

The studies on neuronal activity of the cerebral cortex performed in my laboratory in the Department of Physiology, University of Helsinki have all been done using stumptail monkeys (*Macaca speciosa*). This species of monkey is a close relative of the rhesus monkey *(Macaca mulatta)*; most of the anatomy and cortical cytoarchitecture of the rhesus monkey also applies to the stumptail. However, this species is known for its docile behaviour in the company of man; an advantage over the ferocious rhesus monkey in long laboratory experiments (Kling and Orbach 1963). However, the supply of stumptail monkeys in their natural habitat in South-East Asia is getting sparse; consequently these monkeys are nowadays

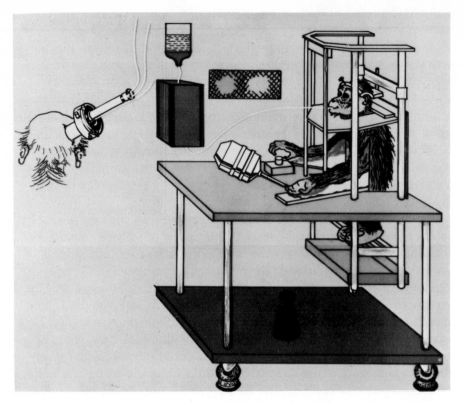

Fig. 10. A stumptail monkey sitting in a specially designed training and recording apparatus. The two signal lights described in the text (p. 30) are shown in the background. The monkey is sensing a vibrator tip with its left hand and pressing the response button with its right hand. The head is fixed with a halo device in the framework of the chair, and a hydraulic microdrive is positioned above the target area (*inset*). (Hyvärinen et al. 1974)

quite difficult to obtain (Marx 1973). We have, therefore, maintained a small colony of them (15 to 20 monkeys) (Hyvärinen et al. 1978c) and managed to breed them with reasonable success (Hyvärinen and Linnankoski 1979).

The recording methods have been described in detail in the original reports. All the studies have been performed with nonanaesthetized, conscious animals capable of moving themselves within the confines of the specially designed chair in which the animals sit during the experiment (Fig. 10). The recordings performed in non-anaesthetized animals require that due attention is given to the comfort of the animals during the experiments. All necessary operations are done under general anaesthesia one or more days prior to the recordings.

The methods used in the recordings are *adaptations of similar methods used in the care and study of human patients.* This ensures reasonably that the recording procedures are tolerated by monkeys as they are by human patients. For stability of recordings the head of the monkey is fixed in the chair using a halo device (Fig. 10) that resembles devices used for the stabilization of the skull or facial bones in humans who have suffered fractures of the head or neck (Perry and Nickel 1959, Prolo et al. 1973). Such stabilization of the head is well tolerated by human patients for several weeks, and monkeys do not appear much disturbed by the fixation or the halo which they also wear when free in their cages between recordings. The electrical recordings are performed in conscious animals just as they can be performed in conscious humans during stereotaxic neurosurgery. In principle the recording arrangement is a modification of the technique described by Evarts (1968).

The microelectrodes that we use are *glass-covered metal electrodes* made of platinum-iridium or tungsten. Our procedure for making glass-coated tungsten electrodes was described by Poranen and Hyvärinen (1982). The tungsten wire is first etched with electric current in potassium hydroxide and then coated with heated glass in a puller.

The electrodes are taken to the brain through the intact dura mater using an Evarts' (1966) type *hydraulic manipulator.* This manipulator is attached to a base cylinder previously implanted in the skull above the target area of recordings (Fig. 10). The location of the target area is determined using stereotaxic or other measurements of the skull as reference.

When care is taken to minimize all discomfort of the animals during the recording sessions, they will co-operate with the experimenters or perform successfully the tasks designed for them for many hours each day.

B. Movement and Orientation Selective Neurones in SI

The visual cortices of cats and monkeys contain a large number of neurones sensitive to orientation and movement of lines over the receptive fields. In an early study of SI we described a few neurones sensitive to movement along the skin in a certain direction (Mountcastle et al. 1969) (Fig. 11), and more such neurones were described by Schwarz and Fredrickson (1971a) and Whitsel et al. (1972). To investigate the possible complexity of the functional properties of neurones encountered in SI we performed a study using a variety of stimuli that could reveal complex field properties (Hyvärinen and Poranen 1978a).

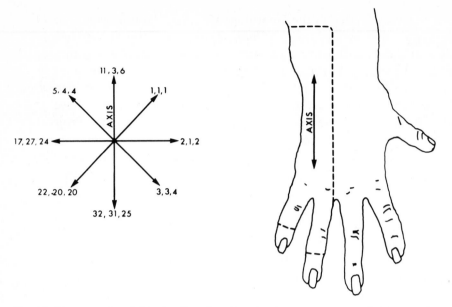

Fig. 11. The receptive field of a directionally selective neurone recorded in area 1 of a rhesus monkey. On the *left* are shown spike counts obtained when a hand-held probe tip moved along the receptive field in different directions. Three consecutive rounds of stimuli were given in each direction. (Unpublished results of experiments performed by Sakata and Hyvärinen in the laboratory of Mountcastle 1968)

Neurones that could not be reliably activated with simple touch on their receptive fields were studied using stimuli moving along the skin. The simplest type of cells *responding to movement along the skin* were those that did not differentiate between different directions of movement or different shapes of the moving stimuli. An example of such a neurone is given in Fig. 12. This neurone gave negligible responses to skin indentation (Fig. 12B) but it responded briskly to movement of the examiner's finger along the skin in any direction (Fig. 12A).

The most common type of neurone not activated with simple skin indentation was the *direction-selective* one. An example is given in Fig. 13. Like the one illustrated in Fig. 12, this neurone responded poorly to puctate stimuli on the skin (Fig. 13B) but it produced a good response to surface movement (Fig. 13A). However, in this case the direction of movement was critical; the cell responded only to stimuli moving distally (Fig. 13A and C).

In area 2 neurones were found that responded to *stationary* or *moving edges* on the skin, the degree of their activity depending on the orientation of the edge. An example of such a neurone is given in Fig. 14.

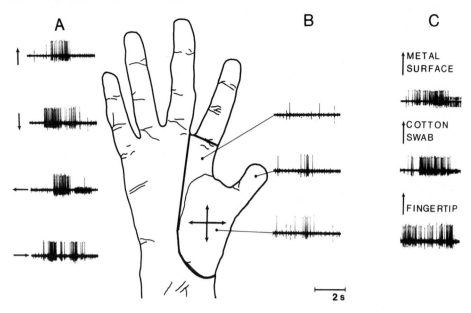

Fig. 12 A-C. The receptive field and the responses of a neurone recorded in area 1; the cell was sensitive to movement along the skin in any direction. The receptive field covered the thenar eminence, the first palmar whorl and the glabrous skin of the thumb. **B** shows that there were negligible responses to punctate stimuli delivered with a probe tip at the loci indicated on the receptive field. **A** shows responses to moving stimuli in four different directions. In **C** the distally moving stimulus was either a metal surface, cotton swab or the examiner's finger-tip all resulting in similar responses. (Hyvärinen and Poranen 1978a)

The receptive field of this neurone covered the first and part of the middle palmar whorl. It responded poorly to stationary or moving punctate stimuli but well to a metal edge placed transversely on the receptive field. In this orientation the response was maximal and it decreased when the orientation was changed (Fig. 14). In different parts of the receptive field the same stimulus orientation produced the maximal response (Fig. 15).

The number of neurones with *complex cutaneous receptive fields increased posteriorly* within SI. Most movement-sensitive neurones that did not differentiate between directions were located in area 1. The more complex cell types specifically sensitive to direction of movement or to orientation of an edge were not observed in area 3b and their number increased posteriorly.

Neurones with complex receptive field properties were relatively rare in SI. Only 6% of the neurones studied belonged to this group. However, the fact that their number increased posteriorly in SI suggests that

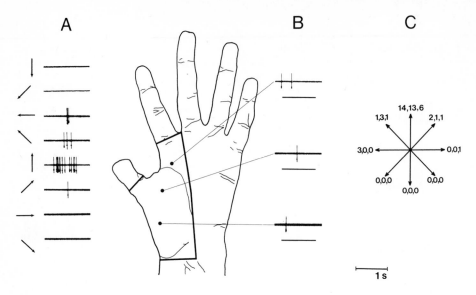

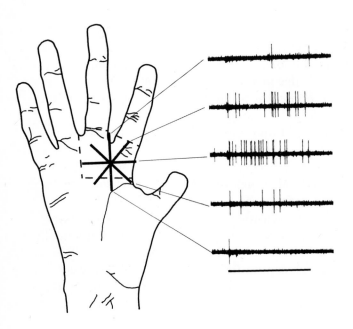

Fig. 13 A-C. The receptive field and responses of an area 1 cell sensitive to distally moving skin stimuli. The receptive field covered the thenar eminence, the first palmar whorl and the base of the thumb. **A** shows responses to punctate stimuli moving in different directions across the receptive field. **B** shows negligible responses to punctate stimuli delivered with the same probe tip at the loci indicated on the receptive field. **C** gives the numbers of impulses evoked on three consecutive cycles of stimuli delivered in different directions across the receptive field. Only the distal-ward direction produced good responses (Hyvärinen and Poranen 1978a)

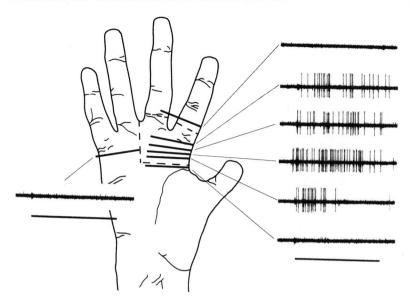

Fig. 15. Responses of the same neurone as described in Fig. 14 to an optimally oriented edge in different parts of the receptive field and outside it. A good response was obtained for an optimally oriented edge in all parts of the receptive field. (Hyvärinen and Poranen 1978a)

Fig. 14. The receptive field and responses of a neurone recorded in area 2. This neurone was sensitive to the orientation of an edge placed on the receptive field which covered the first and second palmar whorls as indicated by the *dashed line*. Across the most sensitive part of the receptive field, between the first and second palmar whorls, a 0.7 mm wide metal edge was positioned in different orientations as indicated in the figure. The metal edge was attached to the tip of a vibrator that produced a steady skin indentation when direct current of 1.3 s duration (*continuous line* under the record) was passed through the vibrator. A good response was obtained only when the orientation of the edge was perpendicular to the axis of the hand. (Hyvärinen and Poranen 1978a)

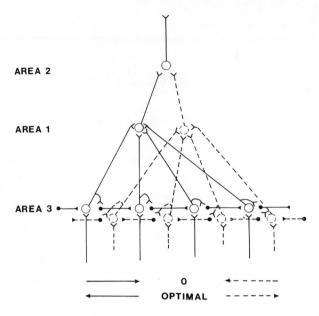

Fig. 16. A hypothetical scheme explaining the directional selectivity of the neurones observed in the primary somaesthetic cortex. Two directionally selective neurones are drawn in area 1; they respond maximally to stimuli moving in opposite directions. The *small black circles* are inhibitory interneurones. The network indicated with *continuous lines* produces optimal responses to stimuli moving to the left and prevents responses to stimuli moving to the right, whereas the opposite is true for the network indicated with *interrupted lines*. The neurone in area 2 would respond to stimuli moving in the two opposite directions, as was observed in a few neurones. The lateral inhibitory connections are not necessarily located in the same cytoarchitectural area but are so presented here for the sake of simplicity. (Hyvärinen and Poranen 1978a)

these properties arise through cortical processing within SI. Directional selectivity is known to be a common property of neurones in the posterior parietal lobe, areas 5 and 7 (Sakata et al. 1973, Hyvärinen and Poranen 1974, Mountcastle et al. 1975, Leinonen et al. 1979), and it is possible that this property arises through further intracortical processing in the posterior association cortex.

Many different anatomical schemes could be proposed to explain the complicated receptive fields. We suggested one possible mechanism (Hyvärinen and Poranen 1978a), illustrated in Fig. 16, which is a modification of the scheme of Barlow and Levick (1965) for explaining directional selectivity in the visual system. Obviously many other schemes could be devised to explain how the excitation produced by movement in one direction reaches the neurone, whereas excitation caused by movement in the opposite direction is inhibited. Within the visual system the retinal level already contains complicated synaptic connections, and the ganglion cells are two synapses centrally from the receptors. The corresponding synaptic level in the mechanoreceptive somaesthetic system lies in the thalamus or cortex since the first neurones extend all the way to the dorsal column nuclei, the second to the ventrobasal complex of the thalamus and the third to the cortex. Thus it is natural that the receptive field properties of the neurones are in general more complex in the visual than in the somatosensory cortex.

The purpose of the neurones with the complex cutaneous receptive fields could be to aid *active sensing of objects during manual exploration*. During such exploration a synthetic picture of the three-dimensional structure is constructed from the stimuli provided by the edges of the object. This stereognostic capacity is severely impaired after lesions in posterior SI (Randolph and Semmes 1974). The stereognostic function is also aided by sensory feedback of the direction of movement along the skin. Thus neurones demonstrating movement and edge sensitivity as well as orientation and direction selectivity could aid the stereognostic function of the somatosensory cortex and probably operate in connection with the neurones whose activity signals joint position and muscle stretch.

The directionally selective neurones of SI have recently been quantitatively documented by Costanzo and Gardner (1980). Moreover, studying the mechanism underlying direction sensitivity Gardner and Costanzo (1980) showed that stimulation in the non-preferred direction resulted in an inhibitory response in the directionally selective neurones. This finding lent support to the suggestion that an in-field lateral inhibition generates the directional selectivity in SI neurones.

C. Receptive Field Integration and Submodality Convergence in SI

As already mentioned, the thalamic input to the three cytoarchitectural subdivisions of SI, areas 3b, 1 and 2 differ: area 3b receives a heavy thalamic projection from the ventrobasal complex whereas areas 1 and 2 have less dense thalamic projections (Jones and Powell 1970a). Within SI areas 3b, 1 and 2 project to each other, to SII, and further to the posterior parietal association cortex, areas 5 and 7 (Jones and Powell 1969, 1970b, Vogt and Pandya 1978, Künzle 1978). Moreover, neurones related to the skin are more common in the anterior part of SI and those related to joints in the posterior part (Powell and Mountcastle 1959b, Whitsel et al. 1971). Paul et al. (1972), Merzenich et al. (1978), and Kaas et al.

(1979) showed that there are two independent representations of the body surface in SI, one in area 3b and one in area 1. They also found partial evidence for a third representation in area 2. Moreover, selective ablation of the three cytoarchitectural areas of SI leads to differential deficits in somaesthetic tasks: ablation of area 1 in deficits in discrimination of texture, and ablation of area 2 in deficits in discrimination of angles (Randolph and Semmes 1974).

All these studies suggest *functional differentiation within SI in the antero-posterior direction.* Examining the functional types of neurones in the cytoarchitectural subdivisions of the postcentral gyrus we found that the receptive field size and the complexity of the functional properties of the neurones increased posteriorly in SI (Hyvärinen and Poranen 1978b).

In the postcentral gyrus neurones activated by simple touch stimuli on their receptive fields formed the largest group (57% of the neurones studied). The receptive fields of the neurones in the anterior part of the postcentral gyrus were relatively small, and those on fingers were limited to part of one finger (Fig. 18). In the posterior part of the gyrus, however, we found ten neurones that had *discontinuous excitatory receptive fields.* Moreover, all the separate portions of these fields were located in similar regions of adjacent fingers (Fig. 17). For instance the glabrous skin of the distal pad of the second, third and fourth fingers was excita-

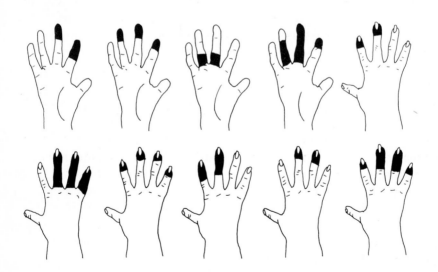

Fig. 17. The receptive fields (*black*) of the ten cortical neurones that had discontinuous receptive fields. Nine of them were located in area 1. (Hyvärinen and Poranen 1978b)

tory for one neurone, whereas the hairy skin of the distal phalanges of the 2nd, 3rd and 4th fingers was excitatory for another. Discontinuous receptive fields were not found on the palm. Nine of these ten neurones were located in area 1.

These kinds of receptive fields probably represent an intermediate phase in convergence towards the large receptive fields that are dominant in the posterior part of the postcentral gyrus. Discontinuous receptive fields have not been observed in the peripheral nerves of the monkey where mechanosensitive receptive fields are continuous, nor have they been observed in the ventrobasal thalamus nor in area 3 of SI. Moreover, it is evident from Fig. 17 that the different receptive fields of one neurone could be innervated by different peripheral nerves. In the primary afferents the fields are small (Talbot et al. 1968), in the anterior SI somewhat larger (Mountcastle et al. 1969), and in the parietal association areas 5 and 7 still larger (Sakata et al. 1973, Hyvärinen and Poranen 1974, Mountcastle et al. 1975, Leinonen et al. 1979). The different fields of area 1 neurones with multiple fields covered corresponding parts in adjacent fingers, i.e., regions that often receive similar stimulation. This finding is reminiscent of the receptive fields described in the second somatosensory cortex (SII) where bilateral receptive fields cover similar distal parts of the body on both sides (Whitesel et al. 1969). It could be assumed that a similarity of function is a cornerstone on which the convergence from different receptive field locations is based in both cases.

In penetrations parallel to the cortical surface in the central sulcus a sequential shift of receptive field locations and dynamic properties of neurones was seen (Fig. 18). During such penetrations the receptive fields of isolated neurones and multiple unit background activity typically shifted around one finger in steps although over distances of about 200-800 μ all the receptive fields had the same approximate location on the skin and were approximately of the same size; the neurones also belonged to the same functional group. Crossing the border from one cell group to another resulted in a change in the location of the receptive fields and often in a change in the functional type of the neurones. The new receptive field location and functional properties of the neurones then prevailed until the next shift took place.

In area 3b the receptive fields were small in comparison with those of neurones in the more posterior subdivisions. In Fig. 19 the receptive fields of 621 neurones are grouped into three categories: small, medium and large. Small ones covered one phalanx or one palmar whorl or less on the skin, one joint or muscle or one well-localized spot in the deep tissues. Medium-sized receptive fields covered two to three phalanges on one finger, or an area of half the palm. Large receptive fields covered se-

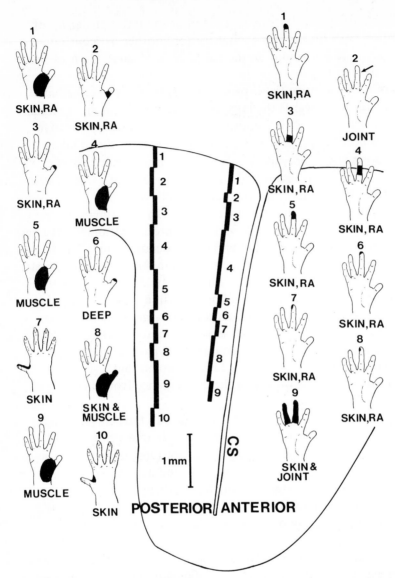

Fig. 18. Reconstructed electrode tracks of two penetrations parallel to the cortical surface and separated laterally by 3 mm in the posterior lip of the central sulcus (*CS*). A shift in the track indicates a change in the receptive field properties of the background activity or of the neurones isolated. The receptive field location and the modality prevailing within the region are marked along the track. (Hyvärinen and Poranen 1978b)

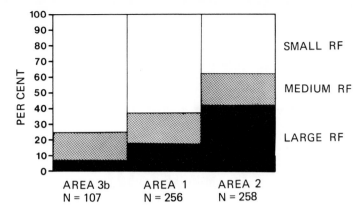

Fig. 19. Percentages of small, medium and large receptive fields (*RF*) in the three cytoarchitectural subdivisions of the hand area of the postcentral gyrus. (Hyvärinen and Poranen 1978b)

veral fingers, the whole palm or an even larger area, or several muscles or joints. In the anterior part of the gyrus neurones with small receptive fields constituted 75%, whereas only 7% of the cells had large receptive fields. The proportion of medium-sized receptive fields was 18%-20% in all areas.

In area 1 the small receptive fields constituted 63% and the large ones 17%. This area also contained more of the neurones that had complicated receptive fields (Table 1). In area 2 the number of neurones activated by simple cutaneous touch had decreased to 42% with a corresponding increase in the number of deep sensory neurones, complicated skin neurones and neurones with submodality convergence (Table 1). All the neurones that responded maximally to an edge in one particular orientation were located here as well as the majority of the direction selective neurones. The receptive field size further increased here (Fig. 19); the proportion

Table 1. Distribution of functional cell types in the cytoarchitectural subdivisions of the hand area of the postcentral gyrus. 632 cells. Percentages are calculated from all the cells recorded in each subdivision. (Hyvärinen and Poranen 1978b)

Cell type	Area 3b		Area 1		Area 2	
	N	%	N	%	N	%
Simple skin neurones	84	69	164	67	112	42
Deep sensory neurones	15	12	31	13	46	17
Complicated skin neurones	4	3	25	10	33	13
Neurones with convergence	7	6	16	6	48	18
Other cell types	12	10	10	4	25	9
Total	122	100	246	100	264	100

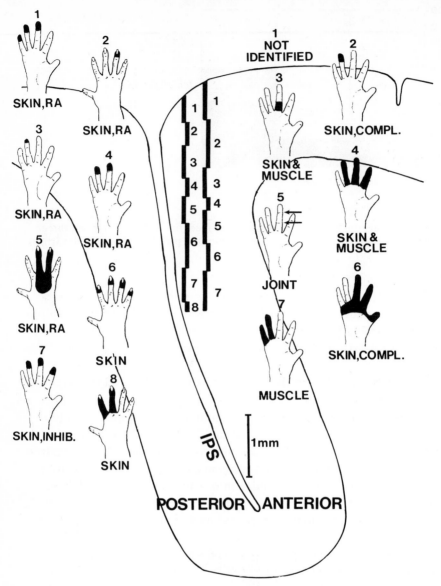

Fig. 20. Reconstructed electrode tracks of two penetrations performed along the cortical surface in the anterior lip of the intraparietal sulcus (*IPS*). The lateral distance between the penetrations was 1.5 mm. The symbols are the same as in Fig. 18. (Hyvärinen and Poranen 1978b)

of small receptive fields was only 38%, whereas the large receptive fields represented 42%.

In penetrations going deep into the intraparietal sulcus the properties of the cell groups encountered shifted similarly as in the posterior lip of the central sulcus. This is illustrated in Fig. 20 which shows results from two penetrations into the posterior part of the gyrus. In tangential penetrations the widths of the regions of cells with similar functional properties were approximately the same as in the anterior part of the gyrus (Fig. 18). However, in the posterior part of the gyrus the penetrations covered larger parts of the hand than in the anterior part of the gyrus. The two penetrations illustrated in Fig. 18 contained neurones with receptive fields on one finger only (thumb and middle finger, respectively), whereas similar penetrations in the posterior part of the gyrus covered most of the hand (Fig. 20). Moreover, posterior penetrations contained more of the complex cell types such as complicated skin neurones and neurones with submodal convergence.

In the tangential penetrations the mean width of the functionally homogenous regions was about 500 μ. This is approximately the width of the cortical area that represents a point on the skin in area 3b in the owl monkey (Sur et al. 1980). The columnar arrangement of neurones that represent different submodalities appears to be organized in antero-posterior bands which were described by Sur et al. (1981) in the owl monkey and the cynomolgus monkey. They found that neurones related to rapidly (RA) and slowly adapting (SA) cutaneous receptors were organized in antero-posterior bands which were continous across the representations of adjacent digits. The sizes of the receptive fields vary, being small for instance in the hand area and large on the back. However, independent of the body part in question, the cortical region representing a region on the skin with the size of a receptive field was constant and approximately 1 to 1.2 mm in diameter, containing one set of adjoining SA and RA bands. The width of the cortical representation of the peripheral input from a region the size of a receptive field was comparable to the width of the "hypercolumns" of the visual cortex (Hubel and Wiesel 1974).

Our studies (Hyvärinen and Poranen 1978 a,b) indicated a functional differentiation in the cytoarchitectural subdivisions of the primary somatosensory cortex. Areas 3b and 1 contained a great number of neurones activated by simple touch, whereas in area 2 an increasing number of neurones with more complicated cutaneous receptive fields was found. Likewise, submodality convergence to single neurones was more common, and the receptive field size increased in the posterior part of the gyrus.

On the basis of anatomical evidence Jones and Powell (1970b) postulated that a chain of converging connections towards the posterior parietal areas was present in the parietal lobe. The projections of area 3, 1 and 2 were studied by Vogt and Pandya (1978), who showed that each of these areas projects to the next one posteriorly but that connections in the opposite direction, towards the central sulcus, were less pronounced. The above findings, supported by the anatomical data, suggest that *an increase in the complexity of the information handled occurs in the cytoarchitectural subdivisions of SI with successive intracortical projection steps from area 3 to areas 1 and 2.* From here the somesthetic information is projected further to the posterior parietal association areas.

D. Influence of Attention on Neuronal Function in SI

A problem of considerable interest in the physiology of cerebral cortex concerns the effect of attention on neuronal function. Previous studies in this field of research have mainly been performed using the evoked potential technique (Hernández-Peón et al. 1956, Hernández-Peón 1961, 1969, Lindsley 1960). However, evoked potentials average the activity of a large volume of neural tissue and do not reveal intracortical mechanisms. In microelectrode recordings the columnar and laminar organization of the cortex can be taken into account. Moreover, in non-anaesthetized, chronically prepared animals, effects of attention can be studied on a cellular level using the transdural microelectrode recording method. When the body parts studied are effectively immobilized, the somatosensory system offers some advantages over the auditory and visual systems in which uncontrollable variations in the pupillary diameter or in the activity of middle ear muscles could influence the results. In the immobilized somaesthetic system variations in the neuronal activity that *correlate with changes in the attentive behaviour of the animal* are results of functional modification of transmission within the CNS. The primary somatosensory cortex is known to respond briskly to cutaneous vibration (Hyvärinen et al. 1968, Mountcastle et al. 1969). We designed, therefore, for the monkey a task that would produce changes in attentive behaviour allowing simultaneous study of neuronal responses (Hyvärinen et al. 1980).

The monkeys were deprived of water in their cages and received their daily ration of drinks during the training or recording sessions. During these sessions one of the monkey's arms was immobilized with plasticene and leather straps in front of the animal in a comfortable position, and on this hand a vibrator tip was positioned (Fig. 10). With the other hand the monkey could press a lever to obtain a juice reward through a pipe mounted close to the monkey's lips. Two signal lights, one yellow and one red, were in front of the monkey at a distance of 40 cm. Figure 21 shows the timing of the task performed by the monkey. An essential part of the stimulus sequence was a cutaneous vibration period lasting 4 s. Prior to the vibration, a signal light (yellow or red) was turned on in front of the

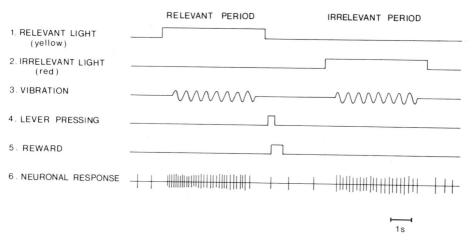

Fig. 21. Part of the continuous sequence of the task of the monkey. The light signals indicated to the monkey whether the vibratory stimulus was relevant or not.
The vibratory stimuli were identical under both conditions. The recording of responses was usually started from the irrelevant phase. The time period from the end
of the irrelevant period to the beginning of the relevant period was the same as between them. (Hyvärinen et al. 1980)

animal. The phase of the task indicated by the yellow light was called
relevant. During this condition the monkey's correct detection of the
end of the vibration period, indicated with a lever press, was rewarded
with fruit juice delivered through an automatic solenoid valve. When the
monkey had learned this part of the task the second, *irrelevant* part of
the task was introduced to it. This phase of the stimulus sequence was
identical with the relevant phase except that the signal light was now red
and no reward was given for the detection of the end of the vibration;
neither was any punishment used.

The monkeys kept a standard of 90% correct performance during the
recordings. When the red light was on the monkeys did not seem to pay
any attention to the cutaneous stimulus but instead chewed food remnants from their food pouches, looked around the laboratory, moved
about in the chair making calls, etc. During the relevant part of the task
the monkeys moved less, and towards the end of the vibration period
they became immobile in order to respond quickly to the termination of
the vibration. Thus there were behavioural indications of attention towards the cutaneous vibratory stimulus during the relevant phase and
signs of inattention towards the stimulus during the irrelevant phase of the
task.

The data were recorded on tape and analyzed with a computer which
constructed two types of histograms of the responses. One of them was a

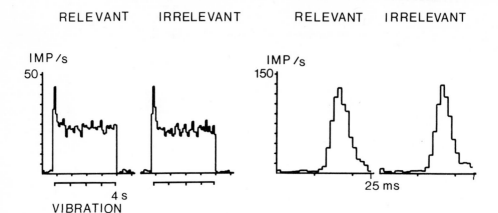

Fig. 22. Responses of a glabrous skin rapidly adapting neurone recorded in area 3b, layer IV. Its receptive field was the volar side of the distal phalanx of the little finger where the vibrator tip was placed delivering 40 Hz vibration with peak-to-peak amplitude of 240 μ. The monkey's performance was 100% correct. (Hyvärinen et al. 1980)

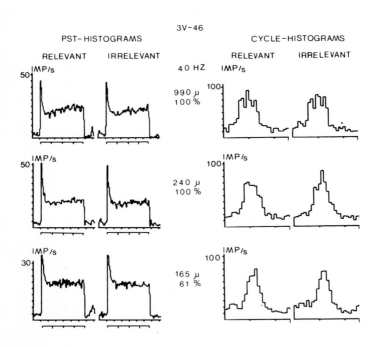

Fig. 23. Responses of a glabrous skin rapidly adapting neurone recorded in area 1, layer IV, to a series of 40 Hz vibration stimuli with a decreasing stimulus amplitude. The level of performance is marked in the *center*. (Hyvärinen et al. 1980)

peristimulus time histogram (PST histogram) in which simple summation of the responses was performed separately for the relevant and irrelevant responses. The other was a *cycle histogram*, which was used as an indicator of the degree of phase-locking of the responses to the sinusoidal stimulus.

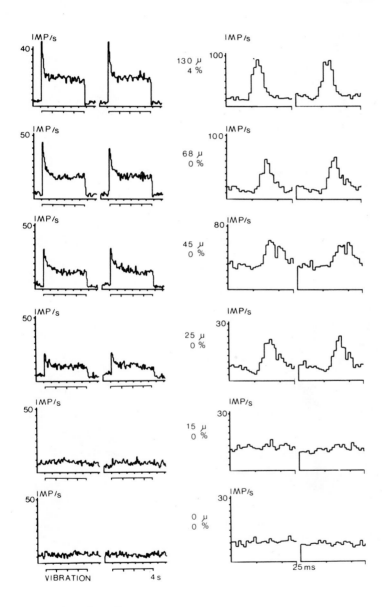

Fig. 23 (continued)

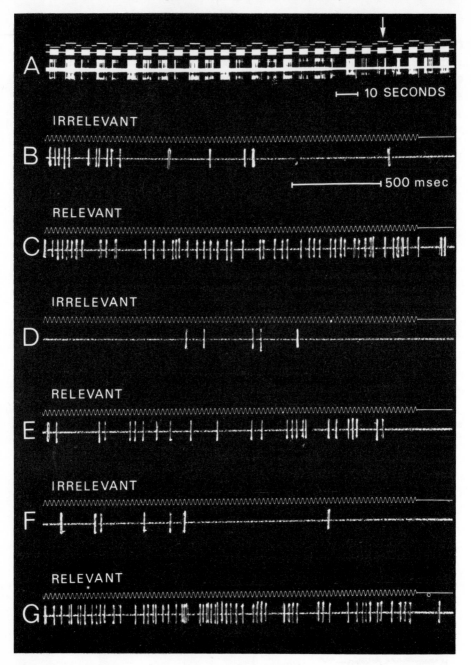

Fig. 24 A-G. Recordings from a cortical neurone during performance of the task with a 40 Hz, 200 μ vibration. **A** Several consecutive responses to relevant and irrelevant vibratory stimuli. The top trace shows the relevant light (yellow) as an upward deflection and the irrelevant light (red) as a downward deflection, the middle position indicating that the lights are off. The middle trace shows vibration; unit responses are seen

Most of the 160 neurones studied in SI during the performance of the task gave clear responses to the vibratory stimuli, but the responses did not differ between the irrelevant and relevant conditions. Figure 22 shows typical PST and cycle histograms for a glabrous skin rapidly adapting neurone recorded in area 3b.

With intensive training the task of the monkey might have become so easy that it could be performed with minimal attention towards the stimulus. Thus, a high level of training could cause the task to become inadequate to reveal the possible effects of attention on cellular responses. The validity of this argument could be tested by reducing the amplitude of vibration until the monkey's level of performance started to fall. When this was done the monkey made numerous errors in the timing of the lever presses. Figure 23 shows a series of recordings made from a rapidly adapting neurone with a decreasing stimulus amplitude. The lowest record shows the control responses when no vibration was delivered. Above it, in a descending order, are histograms indicating responses to successively lower stimulus amplitudes that caused a gradual detriment in the monkey's performance. The intensity of the responses followed monotonically the stimulus amplitude both during the relevant and the irrelevant condition. For the neurones that showed no differences in their responses to the relevant and the irrelevant stimuli, the change in the stimulus amplitude and performance level had no influence on the result. The responses of such neurones remained similar during the relevant and irrelevant phases of the task.

However, there were conspicuous exceptions to this basic finding. In some cells driven well by vibration, a clear difference was seen between the relevant and the irrelevant responses. Figure 24 presents results from such a neurone. In A series of consecutive relevant and irrelevant responses are seen. The vibration was presented at 40 Hz, 200 μ peak-to-peak amplitude. The detection of this amplitude is easy, and the monkey responded correctly to nearly all relevant stimuli. The responses to the relevant stimuli were clearly stronger. At the time marked by the arrow the monkey did not seem to be paying attention to the stimulus, but was

---

on the bottom trace. A difference between the relevant (light trace up) and irrelevant (light trace down) responses is seen in the density of spikes. At the *arrow*, during the relevant stimulus, the monkey did not perform the task but was trying to work food remnants out of the food pouches. **B-G** Latter part of six consecutive irrelevant and relevant responses shown with faster film speed. At the end of all the relevant stimuli the monkey pressed the lever correctly. There are more spikes for the relevant vibration than for the irrelevant. (Hyvärinen et al. 1975)

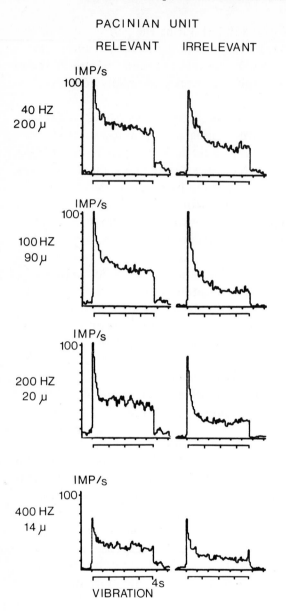

PACINIAN UNIT

RELEVANT   IRRELEVANT

40 HZ
200 μ

100 HZ
90 μ

200 HZ
20 μ

400 HZ
14 μ

VIBRATION

Fig. 25. Responses of a Pacinian neurone recorded in area 1, layer VI, responding to stimulation of the 4th and 5th fingers and the hypothenar eminence. The vibrator tip was placed on the volar side of the proximal phalanx of the 5th finger; the vibration frequencies and amplitudes are marked on the left. The performance level from the top downwards was 100%, 100%, 87% and 64% correct. (Hyvärinen et al. 1980)

trying to get remnants of food from its food pouches. It missed that stimulus and, indeed, the cellular response was much weaker than to those relevant stimuli that it detected correctly. In the lower part of the picture (B-G) the latter parts of three consecutive rounds each of irrelevant and relevant stimuli are presented with a faster film speed. A clearly larger number of spikes is seen in the responses to all relevant stimuli than in the irrelevant responses.

The responses of 16% of the neurones studied were enhanced by the relevance of the task. Figure 25 shows the PST histograms of one such neurone studied at four different frequencies of vibration. At all frequencies tested the responses were greater for the relevant than the irrelevant stimuli.

The effect of the relevance of the task was least common in area 3b; here it was observed in 8% of the neurones tested. In area 1 the effect was present in 22% of the neurones tested. The neurones whose responses were enhanced by attention were unevenly distributed in the horizontal layers of the cortex. Because no lesions were made along the electrode tracks the determination of the layer for each cell was only approximate, the probable error being ± 1 layer. When the percentage of the neurones that were affected by attention was calculated for different layers, *the junctional region between layers I and II, and layer II proved to contain considerably more cells influenced by the task* than did the middle layers (Fig. 26). A slightly larger percentage was again found in layer VI.

In layers I and II of the cortex neurones are not as easily isolated for study with microelectrodes as they are in the middle and deep layers. Therefore, the sample of neurones studied there was smaller than in the deeper layers. However, attentive expectancy had clearly an effect on a larger proportion of neurones in these layers than in the deeper layers. *In the middle layers, particularly in layer IV, the effect was uncommon.* Since the specific thalamo-cortical afferents terminate in layer IV, it is not surprising that the non-specific effects related to attentive behaviour were least common there.

In layer I the most likely cells to be recorded from are the horizontal cells of Cajal, and in layer II, the small pyramidal cells. These cells may receive afferents related to attentive behaviour. The effect of attention may also be a consequence of an integration action of the axonal and dendritic terminals that form the dense neuropil on the cortical surface.

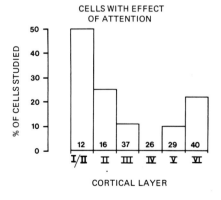

Fig. 26. Distribution of the effect of attention as percentage of neurones studied in different layers of the cortex. The *first column to the left* (I/II) refers to the junctional region between layers I and II. *Numbers in columns* indicate the total number of cells studied in each layer. (Hyvärinen et al. 1980)

The cortical contacts that form local networks have been called *localistic or modular* (Szentagothai 1975), and the cortical modules have been equated with cortical columns (Mountcastle 1978). However, as stated by Szentagothai, in addition to the localistic (modular) type of connectivity there is a more *diffuse interaction* exercised especially over lamina I by cortico-cortical afferents and intracortical axons arising from the deeper layers. These inputs were specifically discussed in our original work on the attention effect (Hyvärinen et al. 1980). On the basis of the uneven distribution of attention effects in the cortical laminae it seems that the incoming sensory information that is received in the middle layers is combined within the cortical modules in the relevant context in the uppermost layers of the cortex where non-specific and association inputs are added to it. *The prominent role of the diffuse interaction system in lamina I is to mediate to the cortical modular system the non-specific excitatory effects triggered by attention.* The combination of attention with the specific stimulus features can thus be realized within the cortical modules. Such a combination appears to be necessary for the conscious perception of the stimulus features and for their storage into memory.

Interestingly neurones responding to attention were concentrated in the superficial cortical layers. However, the lack of attention effects in the middle layers was quite striking, since the activity of over 80% of all the neurones studied was not influenced by attention. Reasonable stability in responses of these neurones could be expected, since many neurones in SI respond well even during general anaesthesia. During explorative manipulation the somaesthetic sense of the hand is used actively to bring the receptive fields of cortical neurones into contact with the stimulus objects. Thus, during active exploration the motor system generates the stimuli which normally modulate greatly the activity of SI. Only by immobilizing the hand it is possible to dissociate the motor component of sensing from a central attentional effect. With the receptive field immobilized, the attention effect should be observable as an augmentation of the responses when the animal or man is anticipating the stimuli and actively attending to them in order to perform correctly a task dependent on the stimulus. However, our results show that this is not true for most neurones in SI. It seems that *the only factor that influences the activity of the majority of neurones in SI is the adequate stimulus.*

In another study, made with multiple unit recording technique during the performance of the same task we investigated other parts of the brain. We found that the *attention effect was considerably more pronounced in the second somatosensory area and in the motor cortex than in SI* (Poranen and Hyvärinen 1982).

Recordings of multiple unit activity were made in SI, SII, the VPL nucleus of the thalamus, and the motor cortex. The multiple unit data were automatically analyzed as so-called "periodograms". These were constructed from the integrated multiple unit activity as summed responses to sine waves of the vibratory stimuli (for closer explanation see Poranen and Hyvärinen 1982). The periodograms shown in Figs. 27 and 28 illustrate the findings in SII and MI.

In the VPL nucleus and in the middle layers of SI no clear effects of the relevance of the task were observed. In SII two different kinds of recording sites were found. One was characterized by clearly localized, contralateral receptive fields and the other one by diffuse, bilateral receptive fields. In the former the vibratory responses were moderately affected by the relevance of the task (Fig. 27A). In the latter the responses to vibration were clearly stronger during the relevant part of the task (Fig. 27B).

In the motor cortex (area 4) synchronous responses to 40 Hz vibration were also observed. These responses were not as clear as in SI but resembled those recorded in SII. They were strongly modified by attention towards the stimuli being enhanced prior to the correct performance of the relevant task (Fig. 28).

Only minor effects of attention were observed in those parts of the somatosensory system that are traditionally considered sensory, i.e., the ventrobasal complex of the thalamus and the middle layers of SI. However, major attention effects were common in SII and the motor cortex. This result recalls those obtained in studies of cortical circulation during sensory and motor tasks in humans. Ingvar (1975), and Ingvar et al. (1976) reported that sensory stimuli enhance circulation in the precentral area, whereas performance of simple movements enhances circulation in the postcentral area. These results indicate that the sensory feedback from the receptors activated during movement could be important in activating the sensory cortex. Attention towards a sensory stimulus, on the other hand, is reflected in the activity of the motor cortex that prepares for a response to the stimulus. These findings emphasize the role of sensory perception in connection with active movements. During normal attentive behaviour the hand is used actively for explorative manipulation. However, when the hand is immobilized, the role of SI may be reduced and SII and the motor cortex may participate in the analysis of the signals   attended to. The enhancement of circulation in the postcentral area during movement may be due to the sensory perception of the movement. *These findings emphasize the role of active movement in activating the sensory systems of the brain.*

A **S II**   A = 990 μ N = 17

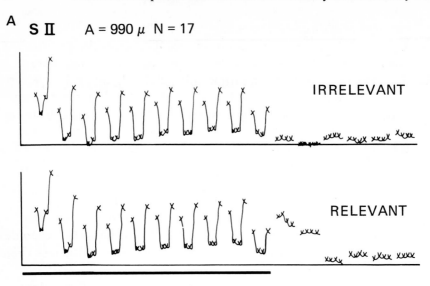

B **S II**   A = 990 μ N = 20

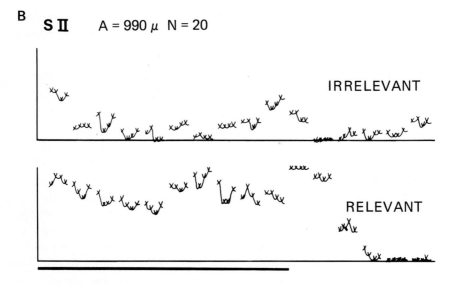

Fig. 27 A,B. Periodograms of multiple unit responses to vibration recorded in SII. **A** Periodogram from a recording site in SII that had a contralateral receptive field. There is little difference between relevant and irrelevant responses, which both show clearly the periodic nature of the response to vibration. **B** Periodogram of a recording site in SII that had bilateral receptive fields. The level of acitivity in the relevant response is considerably higher than in the irrelevant response, but no clear periodicity is observed. (Poranen and Hyvärinen 1982)

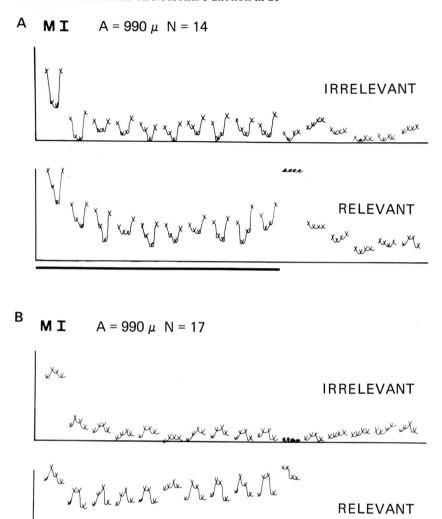

Fig. 28 A,B. Periodograms of multiple unit vibration responses in the motor cortex. **A** Recording site close to the central sulcus in area 4. The periodicity is observable and the relevant response is somewhat stronger than the irrelevant. **B** Recording site 2.5 mm anteriorly from the central sulcus in area 4. The relevant response is much stronger than the irrelevant, but the periodicity is fairly weak. (Poranen and Hyvärinen 1982)

The sensory-motor paradox or the influence of sensory tasks on the motor structures and the influence of motor tasks on the sensory structures may be related to the fact that sensory stimuli that are attended to usually also serve as a basis of a behavioural response. Thus it may be difficult to draw a sharp distinction between the sensory and motor aspects of the neuronal responses since during free behaviour they greatly influence each other.

IV. Neural Connections in the Posterior Parietal Lobe of Monkeys

The neural connections of the posterior parietal associative cortex are complex. In this respect the posterior parietal lobe differs radically from the primary somatosensory cortex in the anterior part of the parietal lobe which receives its predominant input from the ventrobasal complex of the thalamus. No single input dominates the connections to the posterior parietal areas 5 and 7; they receive a multitude of connections from several cortical and subcortical sources. Likewise, the parietal associative cortex projects its efferent connections to numerous different regions.

Summaries of the cortical and subcortical connections of areas 5 and 7 are presented in Tables 2 and 3. As these tables indicate, *the connections of the posterior parietal cortex extend to approximately 60 identified cortical areas or subcortical nuclei,* not counting the connections to the opposite hemisphere which are also numerous. The abundance of the direct neural connections of the posterior parietal lobe suggests that this region is functionally equally complex. The connections of the posterior parietal cortex are briefly summarized here; a more thorough review of them was recently published elsewhere (Hyvärinen 1982).

A. Connections of Area 5

A major thalamic input to area 5 arrives from the *lateralis posterior (LP)* nucleus; the *central lateral (CL)* and the *pulvinar* nuclei also project here (Chow and Hutt 1953, Pearson et al. 1978). The ipsilateral *first and second somatosensory regions* (Jones and Powell 1969, 1970b, Pandya and Kuypers 1969), the *motor (area 4)* and *premotor (area 6)* cortices, and *area 7* project to area 5 (Pandya and Kuypers 1969, Künzle 1978). The *contralateral areas 5 and 7* also project here (Petras 1971).

The efferent cortical connections of area 5 extend to the *ipsilateral SI, areas 7 and 19, premotor cortex,* regions of the *principal and arcuate sulci,* the *cingulum,* the *insula* and the *superior temporal gyrus,* and *contralaterally to SI* and *areas 4,5, and 7* (Peele 1942, Pandya and Kuypers 1969, Jones and Powell 1970b, Petras 1971, Chavis and Pandya 1976). Subcorti-

Table 2. *Ipsilateral cortical connections of posterior parietal lobe of the monkey.* (Hyvärinen 1982). ↑ = connections to the parietal area, ↓ = connections from the parietal area

Area of projection	Parietal area					
	Area 5	Area 7	Area PG	Area PF	Area PGa	Area POa
1. Somatosensory cortices						
SI	↓	↓				↓
Area 3a	↑					
Area 1	↑	↑				
Area 2	↑	↑		↑		
SII	↑			↑		
2. Visual cortex						
Area 18 (OB)		↑	↑			↑↓
Area 19 (OA)	↓	↑↓	↑		↑	↑↓
3. Posterior parietal cortex						
Area 5		↑↓	↑			↓
Medial parietal area			↑			
Area 7	↑↓		↑			↓
7a (PG)	↑			↑↓	↑	
7b (PF)	↑		↓		↑	↑
PGa			↓	↓		
POa	↑	↑				
4. Insular cortex	↓					
5. Temporal cortex						
Area Tpt		↑				
Area 22		↑↓				
Area 20, 21	↓	↓				
Areas 36, 38 (TF, TH)	↑	↑↓	↑			
STS	↓	↓	↑↓	↑↓		↑↓
6. Frontal cortex						
Area 4 (motor)	↑	↑↓		↑		↓
Area 6 (premotor)	↑	↑↓				
Granular (areas 8-12, 45, 46)	↑↓	↑↓	↑↓	↑		↑↓
7. Limbic system						
Cingulum (areas 23-24)	↓	↓	↑↓	↑		
Retrosplenial area			↑			
Hippocampus			↑			
Presubiculum			↓			
Area 27			↓			

Table 3. *Subcortical connections of the posterior parietal lobe of the monkey.* (Hyvärinen 1982). ↑ = connections to the parietal area, ↓ = connections from the parietal area

Area of projection	Parietal area			
	Area 5	Area 7	Area PG	Area PF
1. Basal ganglia				
Caudate	↓	↓	↓	
Claustrum	↓	↓	↑	
Putamen	↓	↓		
2. Thalamus				
Anterior			↑	
Posterior			↑	↓
Pulvinar		↑	↑	
Oral	↑			↑
Medial	↑	↑		↑
Lateral	↑		↑	
Inferior	↑			
LP	↑↓	↓		
LD		↓		
VL	↓	↓		↑↓
VA		↑		
PC		↑	↑	
CM		↑		
Pf		↑		
CL	↑	↓		
MD	↓			
VB	↓	↓		
Reticular	↓			
Zona incerta	↓	↓		
H$_2$ of Forel	↓	↓		
Subst. innominata		↑	↑	↑
3. Hypothalamus		↑		
4. Brain stem				
Colliculus superior	↓	↓	↓	↓
Pontine nuclei	↓	↓	↓	
Raphe		↑	↑	
Locus coeruleus		↑	↑	
Central grey		↓		
Tegmentum	↓			↓
Trigeminal sens. n.			↓	↓
Pretectum	↓		↑	
Substantia nigra	↓			
Dorsum of mesenc.			↓	
Bulbar retic. form.			↓	
Vestibular nuclear c.			↓	
5. Spinal cord				
Pyramidal tract	↓			

cally area 5 projects to the *basal ganglia* (Petras 1971), and to the *thalamic and subthalamic nuclei* indicated in Table 3. Projections to the *brain stem* and the *pyramidal tract* are also strong.

B. Connections of Area 7

Tables 2 and 3 give the connections of area 7 as a whole and separately the connections of subparts of the inferior parietal lobule (areas PG, PF, PGa, and POa), because in many of the older studies no distinction was made between the subregions of area 7.

The main afferent cortical connections to area 7 derive from *ipsilateral SI, SII, area 5,* the *circumstriate belt* and the *frontal cortex,* and the *contralateral areas 5 and 7* (Peele 1942, Jones and Powell 1970b, Petras 1971, Divac et al. 1977). The various *thalamic* and *brain stem* nuclei indicated in Table 3 also project to area 7.

Main efferent cortical connections from area 7 extend ipsilaterally to the *frontal lobe,* the *cingulum,* the *pre- and postcentral gyri, area 5,* the *temporal gyri,* the *superior temporal sulcus (STS), area 19* and the *fusiform* and parahippocampal gyri (Peele 1942, Pandya and Kuypers 1969, Petras 1971, Chavis and Pandya 1976). *Contralateral* connections extend to *areas 1,2,5 and 7* (Peele 1942). Subcortically area 7 projects to the *basal ganglia* and the *thalamic* and *brain stem nuclei* indicated in Table 3. References to these connections and others mentioned in Tables 2 and 3 are listed elsewhere (Hyvärinen 1982).

C. Summary of Connections

Thus the posterior parietal areas have their main connections, most of which are reciprocal, with *sensory cortical regions* (somatosensory and visual), a large part of the *frontal lobe,* the *superior temporal sulcus,* the *cingulum,* the *homologous areas in the opposite hemisphere,* the *basal ganglia,* the thalamic *pulvinar, LP* and *VL* nuclei, the *superior colliculus,* and with the *pontine nuclei.*

There are some interesting differences in the connections of different subregions of the posterior parietal cortex. Area 5 has strong connections with area 2 of the primary somatosensory cortex, whereas area 7 has connections with the associative visual areas. The main thalamic input to area 5 comes from the LP nucleus, but area 7 is more strongly connected with the pulvinar. Area 7b (PF) receives input from the second somatosensory cortex and from the motor cortex, and is reciprocally connected with the VL-nucleus of the thalamus. Area 5 projects strongly to the pontine nuclei and to the pyramidal tract whereas such connections are weak or absent for area 7.

A functional interpretation of the main connections of the posterior parietal lobe suggests that area 5 deals with somatic functions which are also dealt with in area 7b whereas area 7, as a whole, has functions related to vision. The abundance of connections to the prefrontal areas and the cingulum suggest a role in emotional and attentive aspects of behaviour. The connections with the motor and premotor cortices and the basal ganglia suggest a role for the posterior parietal cortex in motor programming, whereas the connections to the trimodal projection zone in the superior temporal sulcus suggest that the posterior parietal cortex participates in high level integration of sensory functions.

V. Symptoms of Posterior Parietal Lesions

The symptoms of damage of the posterior parietal lobe are important for the understanding of the functions of this part of the brain. Therefore, this section, based on a previous article (Hyvärinen 1982), reviews the literature on the effects of such lesions on human patients and monkeys.

A. Humans

Table 4 lists the symptoms that commonly occur in human patients after posterior parietal damage. The symptom complexes differ appreciably in cases of *unilateral and bilateral lesions.* In the early part of this century two important reports of the bilateral syndrome appeared. In 1909 Balint described a patient with bilateral posterior parietal damage caused by cerebral vascular injuries. The patient had difficulties in visual perception of space, size and distance, and in reaching with the right hand under visual guidance. In 1918 Holmes published a study of six World War I veterans wounded symmetrically in both posterior parietal regions. The visual fields

Table 4. Symptom complexes of posterior parietal lesions in humans and monkeys

Symptom	Man	Monkey
1. Visuo-spatial disorientation	+	+
2. Defects in eye movements	+	+
3. Misreaching	+	+
4. Unilateral neglect	+	+[a]
5. Somatic deficits	+	+
6. Finger agnosia	+	
7. Dyscalculia	+	
8. Reading and writing disorders	+	
9. Constructional apraxia	+	
10. Apraxia for dressing	+	

[a] In man neglect is a typical parietal symptom whereas in the monkey it is typical of frontal damage

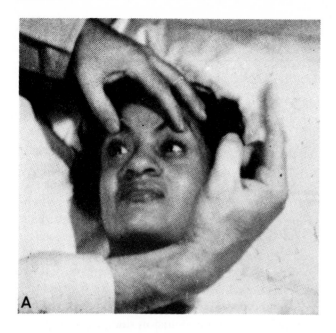

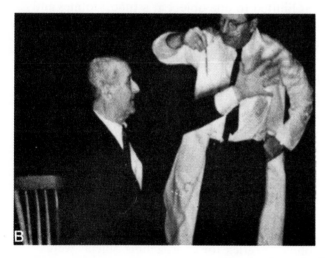

Fig. 29. A Fixity of gaze produced by a biparietal vascular lesion. The patient was unable to relinquish fixation on the examiner when asked to look at fingers on her left side. **B** Misreaching of a patient with biparietal tumour. When the patient was asked to touch the pencil he would usually reach beyond it, although other evidence indicated that he could see it adequately. (Cogan 1965)

of these patients were minimally affected but severe defects were present in visual orientation, judgment of size and distance, visual localization of objects, manual reaching, and voluntary eye movements.

The syndrome described by Balint (1909) consists of three characteristic features: (1) *psychic paralysis of regard,* (2) *optic ataxia* (of the hand under visual guidance), and (3) *difficulty in spatial attention.* With the psychic paralysis of regard Balint meant the patient's inability to look towards a point in his peripheral visual field (Fig. 29). Balint used the term "psychic" to indicate that simple, automatic eye movements were normal in his patient. In analogy with the sensory ataxia of the then prevailing tabes dorsalis Balint coined the word "optic ataxia" to indicate the patient's inability to direct his hand towards visually presented targets. The disturbance in spatial attention was evident in the patient's preference to note visual stimuli presented 35 $^\circ$ to 40 $^\circ$ to the right although he had no visual field defect. It appeared that he could only observe one object at a time and had to be instructed specifically to read each letter in a row. If not instructed he noted only the letter in a row which was about 35° to 40° to the right or the rightmost one in a row. Balint drew the conclusion that the field of attention of his patient was limited to one object at a time. Holmes and Horrax (1919) described a similar defect. Later Wolpert (1924) called the inability of a patient to observe several phenomena simultaneously "simultanagnosia", and Luria (1959), and Luria et al. (1963) described two patients with this disorder.

All *the six patients of Holmes* (1918) who were wounded by bullets bilaterally in the posterior parietal areas presented similar symptoms. The most conspicuous symptom was the inability to seize or touch directly objects which, however, were clearly seen. The patients had a disturbance in orientation and localization in space by sight, and they were unable to estimate absolute and relative distances, lengths and sizes. Another group of symptoms concerned eye movements and ocular reflexes. These patients had difficulty in fixating objects seen and in keeping the fixation while the objects moved; they failed to converge and accommodate for nearby objects, and their blinking reflex for approaching objects were absent.

Several reports of isolated cases of bilateral posterior parietal lesions have since appeared in the literature (Hécaen et al. 1950, Cogan and Adams 1953, 1955, Hécaen and de Ajuriaguerra 1954, Saraux et al. 1962, Godwin-Austen 1965, Tyler 1968, Allison et al. 1969, Eyssette 1969, Kase et al. 1977).

The *syndrome of unilateral posterior parietal damage,* usually of the right hemisphere, was described more recently. This syndrome was called *"amorphosynthesis"* by Denny-Brown et al. (1952), Denny-Brown and

Banker (1954), Denny-Brown and Chambers (1958) and *"apractognosia"* by Hécaen et al. (1956). Whereas the symptoms of bilateral damage affect both sides equally, the unilateral syndrome affects sensory experiences and actions mainly on the side contralateral to the damage. Thus a lesion restricted to one posterior parietal region produces a deficit called *"extinction"*. This means that the patient does not notice that one of two *rivalling* (simultaneously and symmetrically presented) *sensory stimuli*, somaesthetic or visual, which is presented on the side opposite to the lesion (Denny-Brown et al. 1952). Denny-Brown and Banker (1954) have shown that a lesion in the left parietal association area can also produce this syndrome in a right-handed person. The right side appears more specialized in spatial analysis and motor praxis but damage to the left side may also occasionally disturb these functions, although damage to the left side more often affects mechanisms of language.

1. Visuo-Spatial Disorientation

On the basis of his studies of eight patients with bilateral posterior parietal gunshot wounds Holmes listed the following *visual disorientation symptoms* produced by such lesions (Holmes 1919, Holmes and Horrax 1919):

1. An error in absolute localization of objects evident as an error in touching them and accurately pointing to them.

2. Errors in relative localization of several objects. The patient could not differentiate which one of several objects was nearest and which farthest, which one most to the left or to the right.

3. Inability to compare the sizes of objects.

4. Difficulty in avoiding obstacles when walking and difficulty in finding one's way.

5. Difficulty in counting objects.

6. Impairment of visual attention.

7. Inability to recognize movement in a sagittal plane.

8. Disorders of ocular movement including so-called fixity of gaze.

9. In one patient loss of stereoscopic vision (Holmes and Horrax 1919).

These defects were *specifically visual*; the patients localized normally sound and touch. Similar symptoms affecting only the contralateral homonymous visual half-fields have been described by many authors in patients with unilateral posterior parietal damage (Riddoch 1935, Brain 1941, Paterson and Zangwill 1944, Bender and Teuber 1947, Critchley 1953, Ettlinger et al. 1957, Ratcliff and Davies-Jones 1972).

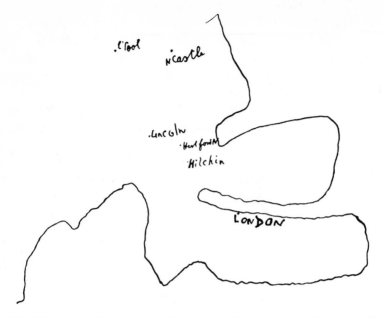

Fig. 30. A map drawn by a patient who had a right parietal tumour. He was asked to draw a map of Englang and to insert any six towns. Note omission of the west coast (the left side). (Critchley 1953)

The patient may also be *unable to observe an object as a whole* by looking at one detail after another. For instance, he may see and recognize a lighted cigarette but not the person smoking, or recognize the window-latch but not the window itself. Difficulties in spatial apprehension are also reflected in reading or drawing maps. The patient may draw a grossly distorted picture of a familiar country and place important cities in wrong locations (Fig. 30) (Critchley 1953). Because of the difficulty in forming visual images of topographical or geographical type, many of these patients often find themselves lost (Brain 1941, Critchley 1953, Whitty and Newcombe 1965). Semmes et al. (1963) have suggested that the difficulties in orientation may be related to lack of attention to the visual background of surrounding stimuli. However, tactually presented maps are also inadequately understood (Teuber 1963) and tactile search is impeded (De Renzi et al. 1970). When the disease affects the dominant parietal lobe interpretation of pictures may also suffer. The theme of the picture — what people are doing and where the scene takes place — is not understood (Critchley 1953). This symptom is related to the inability to observe several objects at one time, the simultanagnosia of Wolpert (1924) and Luria (1959), Luria et al. (1963) called "piecemeal perception" by Paterson and Zangwill (1944).

2. Defects in Eye Movements

Bilateral lesions in the posterior parietal cortex cause eye movement disturbances (Table 5) documented abundantly in clinical literature (Balint 1909, Holmes 1918, Holmes and Horrax 1919, Paterson and Zangwill 1944, Cogan 1953, 1965, Cogan and Adams 1953, 1955, Hécaen and de Ajuriaguerra 1954, Altrochi and Menkes 1960, Luria et al. 1963, Godwin-Austen 1965, Tyler 1968, Allison et al. 1969). Although spontaneous and reflexive movements of the eyes are normal in these patients as well as eye movements during sleep (Michel et al. 1965), *voluntary eye movements towards targets are grossly abnormal*. The fixity of gaze already mentioned (Fig. 29), slowness in voluntary direction of gaze to targets, and abnormal visual searching movements are conspicuous symptoms. Voluntary fixation of an object in the periphery of the visual field may succeed only after several seemingly randomly oriented eye movements often assisted with head movements. Visual fixation and tracking of slowly moving targets may be disturbed and tracking occur through jerky "cogwheel" movements of the eyes. Defects in accommodation and convergence occur (Holmes and Horrax 1919, Michel et al. 1965), and optokinetic nystagmus may be disturbed (Cogan and Loeb 1949, Critchley 1953, Carmichael et al. 1954, Smith and Cogan 1959, Smith 1963). In unilateral lesions reaction times of visually evoked saccades towards the contralateral visual hemifield may be prolonged (Sundqvist 1979).

These defects in eye movements may be described as *ocular motor apraxia*; they differ from the eye movement disturbance produced by frontal lesions which is characterized by paralysis of conjugate gaze (Cogan 1965).

Table 5. Defects in eye movements after posterior parietal lesion in man

1. Fixity of gaze
2. Inability to voluntarily direct the gaze to targets
3. Abnormal visual search
4. Inability in maintaining fixation
5. Inability of visual tracking of slowly moving targets
6. Prolonged reaction times of visually evoked saccades
7. Defective accommodation and convergence
8. Disturbance in optokinetic nystagmus

Together these defects form an oculomotor apraxia

3. Misreaching

As mentioned earlier Balint (1909) noticed in his patient *a disorder of reaching movements performed with the* (right) *hand under visual guidance*. Balint called this symptom "optic ataxia". Although ataxia is not a proper description of the disorder, this term has frequently been used to describe it (Hécaen et al. 1950, Hécaen and de Ajuriaguerra 1954, Eyssette 1969, Michel and Eyssette 1972, Rondot and de Recondo 1974, Damasio and Benton 1979); occasionally it has been called "optic apraxia" (Critchley 1953). Holmes (1918) also noticed defects in reaching movements performed with either arm. This defect is illustrated in Figs. 29 and 31. Recently Damasio and Benton described the reaching disorder in a patient with bilateral posterior parietal lesions in the following way: "She consistently misreached for targets located in the nearby space, such as pencils, cigarettes, matches, ashtrays and cutlery. Usually she underreached by 2 to 5 inches, and then explored, by tact, the surface path

Fig. 31. A patient with Balint's syndrome attempting to pour fluid into a glass. (Allison et al. 1969)

leading to the target. This exploration, performed in one or two groping attempts, was often successful and led straight to the object. Occasionally, however, the hand would again misreach, this time on the side of the target and then beyond it. Another quick tactually guided correction would then place the hand in contact with the object... In striking contrast to the above difficulties was the performance of movements which did not require visual guidance, such as buttoning and unbuttoning of garments, bringing a cigarette to the mouth, or pointing to some part of her body. These movements were smooth, quick and on target."

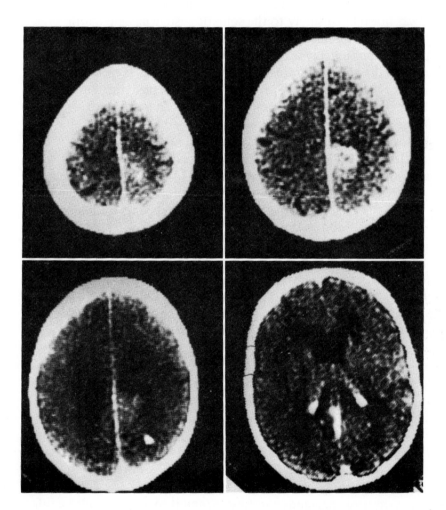

Fig. 32. Computerized tomograms of a tumour in the right superior parietal lobule. The patient's left arm misreached into both visual hemifields but his right arm misreached only into the left visual hemifield. (Levine et al. 1978)

In unilateral lesions the reaching disorder is limited to *reaching to-wards the contralateral visual hemifield* (Riddoch 1935, Brain 1941, Bender and Teuber 1947, Cole et al. 1962, Ratcliff and Davies-Jones 1972, Rondot and de Recondo 1974). In most of these patients the defect has affected reaching with either hand into the contralateral hemifield during fixation of gaze. However, Levine et al. (1978) observed a patient in whom a tumour (Fig. 32) in the right superior parietal lobule caused misreaching with the right arm into the left visual hemifield and with the left arm to both hemifields. When fixating the target the patient misreached only with the left arm and only when not seeing the arm. The visual disorientation symptoms of this patient were limited to misreaching, but the misreaching was not related solely to the somatic system since the arm ipsilateral to the lesion misreached in the contralateral visual hemifield. Thus the defect presumably affected a mechanism which ties the reaching movements to the visual coordinates of the target. The lesion was located in the superior parietal lobule (Fig. 32) in a region called area 7 by Brodmann (1907) but PE by von Economo (1929) which corresponds with area 5 of the monkey. From microelectrode recordings we know that reaching movements are strongly represented in area 5 of monkeys (Mountcastle et al. 1975), and that few neurones in area 5 are activated by visual stimuli (Sakata et al. 1973, Sakata 1975). If the mechanisms of area PE related to reaching are similar in man and monkey, the somatic coordinates of the targets may be given to area 5 neurones on the basis of visual information.

4. Constructional Apraxia

Constructional apraxia is manifested as an *inability to put together one-dimensional units to form two- or three-dimensional figures, patterns or constructions* (Critchley 1953). It may be demonstrated in patients with posterior parietal damage with help of drawing or construction tasks. The patient may grossly simplify the model (Fig. 33), scatter parts of the construction, place the construction close to or on the model (the closing-in phenomenon), and lose track of the vertical and horizontal axes (Hécaen and Albert 1978). In these patients a defect in spatial orientation may become apparent only in a constructive task guided by vision (Paterson and Zangwill 1944, Critchley 1953, Ettlinger et al. 1957, Whitty and Newcombe 1965). Constructional apraxia affects the visual space within the operational sphere of the hands and fingers, and it seems to involve a difficulty in changing the hand from being a part of the person into a tool manipulated in the extrapersonal space (Critchley 1953). Constructional

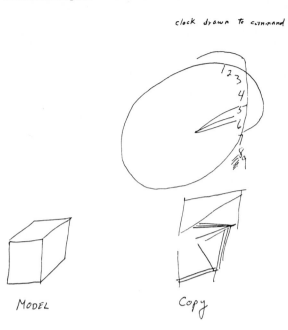

clock drawn to command

Fig. 33. Constructional apraxia evident in drawings of a patient with a right parietal lesion. The numbers and the arms on the clock are crowded on the right side. (Hécaen and Albert 1978)

MODEL   Copy

apraxia thus includes elements of apraxia as well as agnosia; the term apractognosia attempts to express this dual nature (Hécaen et al. 1956, Hécaen 1967, Hécaen and Albert 1978).

Constructional apraxia is most often caused by posterior parietal damage, usually in the right side but occasionally also in the left side or after bilateral lesions. Usually it affects operations performed with either hand. Bilateral lesions may result in both visual disorientation and constructional apraxia (Critchley 1953, Hécaen and Albert 1978). A recent study confirms a special role for the posterior parietal cortex in constructive tasks by showing that the execution of constructional tasks in the extrapersonal space causes a specific increase in the blood flow of the posterior parietal region in humans (Roland et al. 1980).

5. Unilateral Neglect

A common symptom produced by right-sided posterior parietal lesions is *inattention towards the contralateral body half and visual space* (Brain 1941, Roth 1949, Denny-Brown et al. 1952, Critchley 1953, Hécaen et al. 1956). Less frequently this syndrome occurs after left-sided lesions (Denny-Brown and Banker 1954). If the lesion includes part of the temporal lobe the neglect may also include contralateral auditory stimuli

(Heilman and Valenstein 1972a). Patients with a lesion in the right posterior parietal region may completely ignore everything on their left side and behave as if that part of the body and space did not exist (Grüsser 1982). For instance, when a medical examiner speaks to the patient from the left side the patient gives the answer to the right. The patient may dress, comb, shave, or make-up only the right body-half and even state that his left arm or leg is not his but belongs to the patient in the neighbouring bed. This symptom has been called *asomatognosis* or *hemidepersonalization* (Heilman 1979). The patient may show no reaction to painful stimuli applied to the neglected limb (Schilder and Stengel 1928, 1931, Denny-Brown and Banker 1954).

In construction or drawing tasks these patients usually omit the left side of the model and use only the right part of the space provided (Fig. 34). A similar defect was also conspicuously present in the self-portraits of the German artist Anton Räderscheidt whose case Jung (1974) has admirably documented (Figs. 35–37). Figure 35A shows a self-portrait painted prior to the disease and Fig. 35B an attempt to paint a self-portrait two months after a right-sided posterior parietal lesion caused by a vascular accident. The neglect of the left side of the painting is obvious. At first glance it is not so obvious why the right body half is missing. However, considering the technique of painting self-portraits by looking in a mirror it becomes evident that the missing body half is actually the left one, i.e., the one contralateral to the lesion. Thus the lack of the left side of the painting is a consequence of summation of the visual and somatic neglects.

Fig. 34. An example of hemispatial neglect. On the *left* a daisy drawn by the examiner; the *right* one was drawn by the patient. (Heilman 1979)

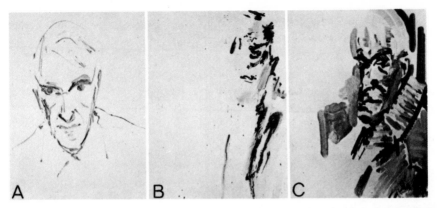

Fig. 35 A-C. Self-portraits of the German artist Anton Räderscheidt before and after a thrombosis of the right posterior branch of A. cerebri media. **A** A self-portrait two years before the stroke. **B** The first attempt to draw a self-portrait two months after the stroke. **C** Another self-portrait made five months after the stroke. (Jung 1974)

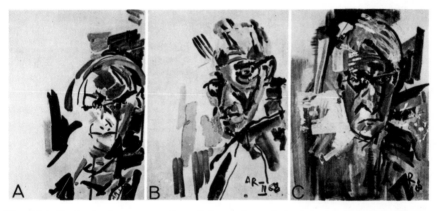

Fig. 36 A-C. Gradual compensation of the neglect of the left side in self-portraits of Anton Räderscheidt. **A** 3 months, **B** 6 months, **C** 9 months after the stroke. (Jung 1974)

Figure 35C shows another self-portrait made five months after the stroke. It shows improvement over Fig. 35B, but the left side of the figure is still rudimentary. Figure 36 shows further increase in detail on the left side resulting from recovery during the subsequent half year. However, even in the final version painted eight months after the stroke the left side is less distinct than the right side, as is evident from the two halves separately replicated as mirror images (Fig. 37A and B).

The somatic neglect often includes various forms of *unawareness or denial of the illness (anosognosia)*, although the patient may suffer from

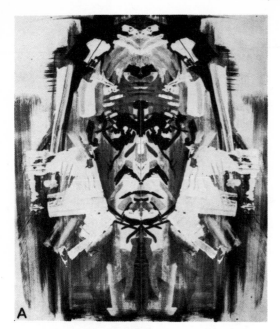

Fig. 37. A The left side of the self-portrait in Fig. 36c duplicated as a mirror im-mage. **B** The right side of the same figure similarly treated. (Jung 1974)

paralysis of the opposite body half, and illusory perception of the opposite limbs such as undue heaviness or a phantom third limb. In motor functions the neglect is reflected as impairment of voluntary movements on the side contralateral to the lesion without deterioration of muscle power. As Critchley (1953) has suggested, the limbs of the affected side appear to occupy a lower level of hierarchy of personal awareness.

When the neglect symptom is less evident it may not become apparent when stimuli are presented separately on either side. However, if two symmetrically located stimuli are presented simultaneously (rivalling stimuli) the patient may notice only the one ipsilateral to the lesion. The term *"extinction"* is used for the neglect of the contralateral one of two rivalling stimuli (Bender and Furlow 1944, Denny-Brown et al. 1952).

Neglect more commonly ensues from lesions in the right hemisphere (Hécaen 1962). Electroencephalographic evidence by Heilman and Van Den Abell (1980) suggests that the right hemisphere may be capable of attending to both sides, whereas the left hemisphere attends to the right side only. This could explain the dominance of the right hemisphere for attention.

In humans unilateral neglect is most commonly produced by lesions in the posterior parietal cortex, but this symptom may also arise from lesions in the frontal lobe (Heilman and Valenstein 1972b), the basal ganglia (Hier et al. 1977), the thalamus (Watson and Heilman 1979), and the mesencephalic reticular formation (Weinstein et al. 1955). The neuronal connections of the posterior parietal lobe, reviewed in Chapter IV, are strong with the frontal lobe, the basal ganglia and the thalamus, whereas the reticular formation regulates the level of arousal.

The neural connections of the sites producing the neglect symptom led Heilman and Watson (1977a,b), and Heilman et al. (1978) to suggest that *neglect ensues from an interuption of a cortico-limbic-reticular activating loop* similar to the one proposed by Sokolov (1963). Most commonly the neglect syndrome is explained as a defect in attention towards the contralateral side. A defect in attention may not, however, explain all features of this syndrome. Heilman and Valenstein (1979) found that in a line bisection task patients with contralateral neglect performed poorly in the contralateral half of space but well in the ipsilateral half, and that this difference was independent of directing the patient's attention to either end of the horizontally placed line. Heilman and Valenstein suggested that a *reduced tendency for performing motor acts, i.e., hypokinesia*, in the opposite half of space explains the neglect better than a defect in attention towards sensory signals. A defect in memory of stimuli presented on the contralateral side is also present in these patients (Samuels et al. 1971, Heilman et al. 1974) and probably contributes to neglect syndrome.

Bisiach et al. (1979) criticized the interpretation of Heilman and Watson (1977a,b), and Heilman et al. (1978) which explains contralateral neglect solely on the basis of input-output connectivities. Bisiach and Luzzatti (1978) studied two natives of Milan with right-sided posterior parietal damage who failed to recall from memory the left side of the scenery at Piazza Duomo in the center of Milan. The neglected scenery was to the left from the patient's imagined position and shifted with the patient's imagined shifts around the square. Bisiach et al. (1979) also found that patients with right-sided posterior parietal damage neglected the left side of the images that they had reconstructed from stimuli appearing successively in time in a narrow vertical slit. These results indicate that in neglect *the spatial scheme of mental representation of the contralateral side is disturbed.* Such findings suggest a topological relationship between the neural representation and the mental spatial scheme.

The results of Bisiach et al. are highly interesting, but as suggested previously (Hyvärinen 1982), they are not necessarily incompatible with the anatomo-physiological explanation of Heilman and Watson (1977a,b) based on the mutual connectivities of the regions producing neglect. The left side of the reconstructed image may predominantly be stored in memory in the right hemisphere. A right-sided lesion reduces the activity of the right cortico-limbic-reticular pathway. Experiments with computer models show that ascending input from the reticular activating system can enhance storage into memory (Rovamo and Hyvärinen 1976, Kohonen et al. 1981). As Heilmann et al. (1974), and Heilman and Valenstein (1979) indicate, a learning or memory defect could also be brought about by an inefficient reticular activating loop. Thus the results of Bisiach et al. (1979) could be explained on the basis of less efficient memory of the left side of the image.

As this discussion shows, the neglect syndrome concerns the following psychological concepts that are closely related: *attention, intention, orientation, learning, memory, and action.* Thus the neural mechanism damaged in the contralateral neglect syndrome is presumably related to all these functions.

6. Gerstmann Syndrome

Gerstmann (1930) discovered a syndrome produced by lesions in the *angular gyrus of the dominant* (usually left) *hemisphere.* This syndrome consists of four symptoms: *(1) finger agnosia, (2) pure agraphia* (without alexia), *(3) right-left disorientation,* and *(4) dyscalculia.* The predominant symptom of the syndrome is finger agnosia, which is characterized by an inability of the patient to recognize, distinguish and name different fingers on his own hands or on those of other persons. Finger agnosia concerns both hands, and is considered an isolated, localized form of a more general symptom called *autotopagnosia* or inability to localize and name various body parts (Critchley 1953). The right-left disorientation concerns all body parts and may extend to the body parts of other individuals. Only the understanding of the lateral dimensions of space is affected; the patient does not mix front and behind, or above and below. As Critchley (1953) remarked, youngsters, the immature or unintelligent (recruit?) rarely, if ever, confuse other dimensions in space except those which pertain to sidedness. Agraphia may involve a disturbance in the execution of letters or words. The patient can usually copy and apparently read and understand what he reads. Calculation defects are primarily related to inability to differentiate between the numerical categories of units, tens, hundreds, etc. (Hécaen and Albert 1978).

The four elements of the Gerstmann syndrome also occur independently of each other. Thus Benton (1961) observed in a statistical analysis that the likelihood of the individual symptoms of the Gerstmann syndrome occurring together is no greater than the likelihood of their occurence with some other symptoms. Therefore, he questioned the occurrence of the Gerstmann syndrome as an entity. Similar although less critical remarks were presented by Critchley (1966). However, as pointed out by Hécaen and Albert (1978); the statistical interactions of the symptoms of the Gerstmann syndrome with other symptoms does not disprove the existence of the Gerstmann tetralogy, because the existence of isolated cases with these four symptoms suggests a common localizational basis. The finding of a similar developmental syndrome in children (Benson and Geschwind 1970) also supports a common mechanism for the symptoms comprising the Gerstmann syndrome.

The role of the fingers is central in the Gerstmann syndrome. This was recognized by Critchley (1953) who developed this theme in the following way: "Perhaps there is something unusual about the significance of the hand in the body-scheme which is the basis of an isolated finger-agnosia; independent of any other trace of autotopagnosia. The hand plays a peculiar role in human ecology, being part of the anatomy which is usually exposed; which is most of the time within the view of the individual subject; which constitutes the chief and most efficient organ or tool; and which, if not the most delicate sensory discriminative region is, at least, the principal instrument of touch... If man is to be regarded essentially a *homo faber*, then it is the hand with which he uses and directs the implement... The hand is largely an organ of the parietal lobe." (Critchley 1953, p. 210)

B. Monkeys

The defects produced in monkeys by ablation of the posterior parietal lobe are listed in Table 4 together with the symptoms produced by lesions in humans. Several of the symptoms in monkeys are in principle similar to those found in humans, for instance visuo-spatial disorientation, eye movement defects, misreaching, and unilateral neglect, although they are less obvious than in man. However, symptoms related to human species-specific behaviour such as calculation, reading and writing, and constructive work are naturally lacking in monkeys. In addition to true species differences, the difficulty in communication with monkeys may contribute to the differences between humans and monkeys. The defects in humans are usually detected on the basis of verbal communication and the patient's attempts to perform in accordance with instructions. Monkeys have to be observed through less direct communication which often involves experimental conditioning to perform pre-planned tasks.

Variation in the locus and extent of the lesion in the posterior parietal cortex may also contribute to the differences observed in the symptoms in monkeys and humans. In monkeys the lesions have usually comprised most of the parietal cortex often including parts of postcentral, preoccipital, and posterior temporal regions. There are only a few studies on either species in which the lesions have been limited to one cytoarchitectural region only. Therefore, it may very well be that comparable functions have not been studied in humans and monkeys affected by comparable lesions.

1. Visuo-Spatial Disorientation

The impairment of visuo-spatial orientation after posterior parietal lesions is less pronounced in monkeys than in man. Monkeys were deficient in finding their home cages in the animal room after posterior parietal lesions (Bates and Ettlinger 1960, Sugishita et al. 1978). Moreover, they suffered from a general spatial disturbance (Hartje and Ettlinger 1973) and directed their manual search in the dark in the wrong direction (Ratcliff et al. 1977).

The defects in spatial orientation produced by parietal and frontal lesions differ somewhat from each other. These differences have been studied using two different kinds of spatial orientation tasks (Pohl 1973). The location of a spatial target may be referred to another object ("allocentric cue") or to the observer ("egocentric cue"). Such studies have indicated that allocentric location is disturbed by posterior parietal lesions whereas frontal lesions have disturbed egocentric location (Pohl 1973, Ungerleider and Brody 1977). The performance of squirrel monkeys was also disturbed after a parietal lesion in a related task in which the spatial locations of a visual cue and the reinforcement were separated (Mendoza and Thomas 1975). Ungerleider and Brody (1977) found that parietal lesions disturbed the monkey's ability to perceive spatial relations between objects, but frontal lesions did not have this effect. They conclude that the anterior frontal cortex mediates a function which is "superordinate to a spatial factor".

Sensori-motor coordination rather than spatial perception has also been disturbed by posterior parietal lesions (Milner et al. 1977). Petrides and Iversen (1979) did not find defects in the performance of monkeys in landmark tests after posterior parietal ablation. On the other hand, their monkeys had difficulty in a visually guided somatomotor task in which a ring-shaped reward had to be passed along bent wires. However, this defect may be considered a form of visuo-motor apraxia rather than a defect in spatial orientation. It illustrates how closely visually guided manual performance and spatial orientation are related.

2. Defects in Eye Movements

Munk (1881) noticed 100 years ago that the ablation of posterior parietal-anterior occipital cortex of monkeys abolished the blinking reflex to approaching contralateral stimuli and caused difficulties in eye movements. Recently more evidence of such disturbances in monkeys has accumulated. Latto (1978) showed that posterior parietal lesions of monkeys impeded visual search, and Stein (1978) showed that cooling of area 7 caused slowing of eye movements towards the contralateral side. Recently Lynch and McLaren showed that bilateral posterior parieto-occipital lesions impaired the ability of monkeys to track visually a fast-moving target. Saccade latencies were also increased, but the ability to maintain accurate visual fixation of a stationary target was not affected (Lynch and McLaren 1979, Lynch 1980a,b). Unilateral lesions produced a slowing of the slow phase of optokinetic nystagmus in the direction towards the side of the lesion, whereas bilateral lesions produced a slowing in both directions (McLaren and Lynch 1979, Lynch 1980a, Lynch and McLaren 1982). The slow phase of vertical optokinetic nystagmus was also slowed after bilateral lesions (McLaren and Lynch 1980).

These results indicate that the posterior parieto-preoccipital cortex is involved in eye movement control in monkeys, too, and that in this respect it resembles its human counterpart perhaps more than was hitherto supposed.

3 Misreaching

Peele (1944) observed that ablation of posterior parietal cortex resulted in clumsiness of the contralateral limbs. After ablation of area 5 the contralateral foot missed the rope which the monkey started to climb; after ablation of area 7 the contralateral hand groped for targets and was ataxic. Visual placing and grasping were impaired after ablation of area 7b, particularly when the monkeys were blindfolded (Fleming and Crosby 1955). Denny-Brown and Chambers (1958) noted clumsiness of reaching and grasping movements, but the hand was brought to the mouth accurately. Ettlinger and his coworkers (Bates and Ettlinger 1960, Ettlinger and Kalsbeck 1962, Hartje and Ettlinger 1973) observed that the reaching defect produced by posterior parietal ablation affected the contralateral arm independently of the direction of its projection and not, as in humans, only when aimed to the contralateral half of space. In these studies visual fixation was not controlled, whereas in the studies on humans the reaching defect was apparent in the contralateral visual hemifield during fixation of gaze. Hartje and Ettlinger (1973) also showed that reaching with the contralateral arm was even more impaired in the dark than light and suspected a non-specific disturbance of spatial orientation as the basis of the misreaching; this interpreation was shared by Ratcliff et al. (1977). The reaching defect is also related to the difficulties in manual maze tasks caused by disturbancies in visuo-motor coordination mentioned above (Milner et al. 1977, Petrides and Iversen 1979). The performance of a complex latch-box opening task was not disturbed by parietal lesions although frontal lesions disturbed it (Deuel 1977). Defects in placing, hopping, grasping, and tactile tasks have also been associated with the reaching defect (Peele 1944, Fleming and Crosby 1955, Bates and Ettlinger 1960, Moffett et al. 1967).

In monkeys the reaching deficit resulting from unilateral posterior parietal ablation is characterized by slowness of movement, abnormal shaping of the hand and fingers approaching the target, and an error in arm projection towards the lesioned side. These errors occur on both sides of the midline, with and without visual guidance. The deficits are transient and recovery takes place in a few weeks (Faugier-Grimaud et al. 1978, LaMotte and Acuna 1978).

The fairly rapid recovery of function after posterior parietal ablation in monkeys led Stein (1976, 1978) to adopt the technique of reversible local cooling in the studies of posterior parietal lobe. In this technique compensatory processes do not influence test behaviour because between the test periods the tissue studied functions normally. It was also significant that Stein studied separately the effects of cooling areas 5 and 7. Cooling of area 5 caused clumsiness of the contralateral arm that misreached in all directions and exhibited difficulty in grasping. Cooling of area 7 produced different symptoms: the contralateral arm misreached in the contralateral visual hemifield, and further lowering of the temperature led to the appearance of a defect even in the reaching performance of the ipsilateral arm in the contralateral hemifield. Stein's results suggest than area 5 contains mechanisms for the somaesthetic control of reaching accuracy, whereas area 7 contains mechanisms for the visual control of reaching accuracy.

4. Unilateral Neglect

Contralateral neglect is one of the most typical symptoms resulting from posterior parietal lesions in man. In monkeys signs of neglect are also produced by posterior parietal lesions but in them this symptom is less conspicuous than in humans. Peele (1944) observed that ablations of areas 5 and 7 lead to reluctance to use the contralateral limbs and weakened reactions to tactile and painful stimuli. Refusal to use the contralateral arm was also noted by Faugier-Grimaud et al. (1978). Denny-Brown and Chambers (1958) noted reduced responsiveness to contralateral visual stimuli after posterior parietal ablation, but visual neglect of single contralateral stimuli was not observed in studies in which the direction of gaze was not controlled (Ettlinger and Kalsbeck 1962, Latto 1977). When the direction of gaze was controlled single visual stimuli were detected in either half of the visual field (Stein 1978, Lynch 1980a), but during cooling of area 7 the monkey's reaction times to stimuli in the contralateral hemifield were lengthened and these stimuli were often ignored altogether. Thus cooling of area 7 caused neglect of the contralateral visual half-field. Stein (1978) concluded that this deficit did not have the nature of sensory blindness but rather that of a defect in the visual control of movement in the contralateral hemifield.

Although the neglect of single contralateral stimuli is not conspicuous in monkeys after posterior parietal lesions, extinction of the contralateral one of two rivalling stimuli, simultaneously and symmetrically presented, has been repeatedly observed. Ablation of posterior parietal-preoccipital cortex, area 5 or frontal cortex produces extinction of contralateral cutaneous stimuli. Posterior parietal ablations also produce extinction of contralateral visual stimuli; the extinction can be detected when two symmetric fixation targets are simultaneously presented (Lynch 1980a). Lesions involving both posterior parietal and preoccipito-temporal cortices lead to extinction of contralateral visual, somatic, and auditory stimuli (Heilman et al. 1970, 1971).

In monkeys the neglect syndrome is readily *produced by lesions in the region of the arcuate sulcus in the frontal lobe.* Bianchi (1895) described this deficit in monkeys and it has later been repeatedly confirmed (Kennard 1939, Welch and Stuteville 1958, Deuel et al. 1979). Monkeys with frontal neglect syndrome recover in a few weeks. Deuel et al. (1979) have recently shown with the radioactive deoxyglucose technique that the neglect produced by periarcuate lesions is accompanied by widespread ipsilateral subcortical metabolic dysfunction and that the recovery is accompanied by normalization of the subcortical metabolic dysfunction.

Watson et al. (1973, 1974) investigated in monkeys whether *lesions in various parts of the assumed cortico-limbic-reticular pathway* produce the neglect syndrome. The view that lesions in this circuitry cause neglect was supported by their findings, indicating that neglect was produced in monkeys by lesions in the cingulum (Watson et al. 1973) and the mesencephalic reticular formation (Watson et al. 1974). The reticular lesions caused slowing of the ipsilateral EEG to a 3 c/s rhythm reminiscent of petit mal epilepsy. The authors interpreted this EEG-change that accompanied the neglect syndrome as a result of cutting off the reticular input to the thalamic non-specific system which was thus left discharging spontaneously. In cats it has previously been shown that unilateral visual neglect is produced by lesions in the caudate nucleus, midbrain reticular formation (Reeves and Hagaman 1971), and the intralaminar thalamic portion of the ascending reticular activating system (Orem et al. 1973). Moreover, studies in cats indicate a projection from the limbic cortex to the reticular formation (French et al. 1955, Adey et al. 1957). Together with the human data, reviewed in Section A.5 of this chapter, these results support the cortico-limbic-reticular theory of neglect proposed by Heilman and Watson (1977a,b).

Watson et al. (1978) also showed, using monkeys with frontal or reticular lesions, that neglect may be a *defect in intention rather than attention.* Their monkeys were trained to respond to a sensory stimulus with the limb contralateral to the stimulus. After cortical lesions these animals responded normally to contralateral stimuli with the limb ipsilateral to the lesions. This response indicated that the lesion did not block the perception of contralateral stimuli. However, the lesioned animals failed to respond with the contralateral arm to stimuli ipsilateral to the lesion. They thus demonstrated an intentional defect in the use of the limbs contralateral to the lesion.

5. Somatic Deficits

Somaesthetic deficits naturally occur if ablations extend to the first or second somatosensory cortices. However, less conspicuous somaesthetic deficits also occur after posterior parietal ablations. Somaesthetic deficits produced by posterior parietal lesions include difficulties in roughness discrimination (Wilson et al. 1960, Semmes and Turner 1977, Stein 1978), tactile shape discrimination (Wilson 1957, Ettlinger and Kalsbeck 1962, Ettlinger et al. 1966, Moffett et al. 1967, Moffett and Ettlinger 1970, Semmes and Turner 1977), retention of somaesthetic tasks and length discrimination (Pribram and Barry 1956), discrimination of lifted weights (Ruch et al. 1938, Semmes Blum et al. 1950), impairment of limb position sense and release of tactile avoiding (Denny-Brown and Chambers 1958). Wilson et al. (1960) observed that bilateral total posterior parietal-preoccipital lesions did not disturb the previously attained level of performance in difficult roughness discrimination but that these lesions prevented the normal improvement in such tasks. Ettlinger et al. (1966) found that tactile shape discrimination of monkeys was impaired only in the dark, and suggested that this defect was caused by a selective motor retardation which was the origin of the defective performance in tactile tasks in the dark. Tactile placing, grasping and hopping reflexes have also suffered from posterior parietal lesions (Fleming and Crosby 1955, Denny-Brown and Chambers 1958, Bates and Ettlinger 1960, Ettlinger and Kalsbeck 1962).

Separate ablation of area 5 did not impair tactile discrimination (Semmes and Turner 1977), whereas its separate cooling did (Stein 1978). This difference could be related to the recovery from the ablation effect or to spreading of the cooling effect to SI. Ablation of area 5 weakened the reaction to tactile stimuli and affected the leg more than the arm (Peele 1944). It also resulted in impairment of tactile placing, holding, grasping and hopping; these functions were impaired by lesions in area 7b, too (Fleming and Crosby 1955), indicating that area 7b participates in somaesthetic functions.

The impairment of the performance of somaesthetic tasks after posterior parietal-preoccipital ablations appears small in comparison with the effect of ablation of the postcentral gyrus (Peele 1944). Although somaesthetic deficits are produced by lesions in posterior parietal areas these areas do not appear to play a major role in somatosensory discrimination (Semmes and Turner 1977).

Posterior parietal lesions may also cause minor somatomotor symptoms such as slight weakness and hypotonia (Peele 1944, Fleming and Crosby 1955). It is difficult however, to separate such symptoms from contralateral motor neglect.

C. Comparison of Monkeys and Man

The somatosensory deficits observed after posterior parietal lesions are more numerous in monkeys than in man. This difference led Ettlinger (1977) to suggest that the posterior parietal function in monkeys is predominantly somatosensory, whereas in humans it is predominantly visual. Such differences between these two species may be real. However, new studies have shown that monkeys also suffer from visual deficits after posterior parietal lesions and the somatosensory deficits appear less significant than those produced by lesions in the somatosensory areas. The species difference between monkeys and man may thus be only a moderate one.

However, many of the symptoms produced by posterior parietal lesions in humans concern typically human activities such as construction, calculation, finger naming, reading and writing, and dressing; functions that are not represented in the monkeys' brains. The other symptom complexes described in humans: visuo-spatial disorientation, eye movement defects, misreaching, and unilateral neglect all have counterparts in the behavioural syndrome produced in monkeys by posterior parietal lesions. Although the details of these symptoms differ in many ways between these species, there clearly is a partial overlap.

One of the major differences between monkeys and man has been in *the misreaching symptom*. This symptom has previously been described in monkeys as affecting reaching with the contralateral arm in any direction. In humans, however, the misreaching has affected reaching with either arm in the contralateral visual hemifield (Ettlinger and Kalsbeck 1962, Drewe et al. 1970, Hartje and Ettlinger 1973). However, cooling of area 7 in monkeys caused misreaching in the contralateral visual hemifield (Stein 1978), indicating that the visual guidance of the reaching movement may, after all, be organized in a corresponding way in both species. The deficits in eye movements are also clearly demonstrated in monkeys although they are less evident than in humans, who have bilateral lesions.

Another major difference between monkeys and humans concerns the *contralateral neglect*. In humans this symptom is typical after posterior parietal lesions, but in monkeys it has seldom been observed in the detection of single unilateral stimuli (Drewe et al. 1970). However, a related symptom is present in both species: the extinction of the contralateral one of two rivalling stimuli. Thus the posterior parietal lobe appears to play some role in the neglect syndrome in monkeys, too, although this role is of lesser significance than that of the frontal cortex. Because of the dominance of the frontal lobe in the neglect syndrome in monkeys and the parietal lobe in humans, Passingham and Ettlinger (1974) considered that spatial perception is likely to be organized in a different way in these two species.

The periarcuate frontal cortex and the posterior parietal cortex are mutually heavily connected, and neglect symptoms can be produced by damage to either area in man and monkeys. The studies of Heilman and Watson and their collaborators reviewed above suggest that lesions in the cortico-limbic-reticular pathway produce the neglect syndrome in both monkeys and man. The optimal locus for the production of this syndrome differs between monkeys and man probably because this circuitry has developed in somewhat different directions in these two species. The mechanisms related to neglect have apparently shifted partially from the frontal to the parietal associative cortex during the development of the human species. Such a shift presumably had some advantages for the development of the human form of life. Perhaps the partial shift of these mechanisms to the parietal lobe was associated with concomitant changes in the mechanisms of attention and intention, the greater flexibility, conceptualization and abstraction of these functions, and with the development of linguistic information processing in humans.

VI. Electrical Stimulation of Posterior Parietal Lobe

A. Monkey

It is difficult to elicit motor effects by electrical stimulation of the posterior parietal lobe. Weak currents have no observable effects on monkeys under deep pentobarbital narcosis. However, several types of responses have been observed when superficial ether anaesthesia and strong currents or chronically prepared awake animals have been used.

In the last century Ferrier (1876) elicited movements of the leg and foot by stimulation of area 5. The Vogts (1919, 1926) observed that stimulation of area 5a caused combined movements of the arms and legs and stimulation of area 5b turning of the eyes downwards with simultaneous limb movements (Fig. 38). Peele (1944) elicited protraction of the upper limb by stimulating area 5 or 7. Fleming and Crosby (1955) observed movements of the contralateral leg, arm and trunk when area 5 was stimulated.

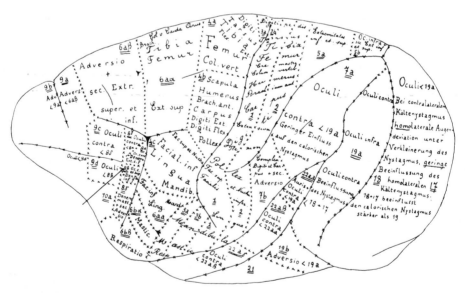

Fig. 38. Motor responses to electrical stimulation of the cortex of *Cercopithecus* monkey obtained by the Vogts (1926)

These studies were all performed under ether anaesthesia. Lilly (1958) stimulated area 5 in non-anaesthetized, behaving monkeys; leg movements were elicited from the medial part and arm movements from the lateral part.

Ferrier (1876) also noted eye movements upon stimulation of area 7. This finding led him to the erroneous conclusion that the visual cortex was located in the angular gyrus, a view rightly criticized by Munk (1881) who, on the basis of ablation studies, placed the visual cortex in the occipital lobe. Von Bechterew (1911) also observed eye movements upon stimulation of the region of the angular gyrus. Stimulation of area 7a produced eye movements, but complex hand and finger movements were elicited by stimulation of area 7b (C. and O. Vogt 1919, 1926). Eye movements elicited by stimulation of area 7 persisted when the pre- and postcentral gyrus were removed, indicating that the posterior parietal cortex had a projection to the oculomotor pathways that bypassed the Rolandic region (Fleming and Crosby 1955). Stimulation of the posterior parts of area 7 led to pupillary constriction and to convergence and accommodation movements (Jampel 1960). In awake monkeys Lilly (1958) elicited both eye and arm

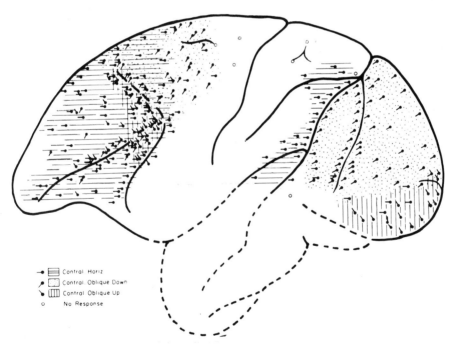

Fig. 39. Cortical map of eye movement responses to electrical stimulation of the alert, cervically transected monkey (*M. mulatta*) obtained by Wagman (1964). Locations of responses are indicated with *solid points*. The *lines* originating from the points indicate the direction of the eye movement. All responses were towards the contralateral side

movements by stimulating area 7. In alert, cervically transected monkeys Wagman (1964) induced conjugate horizontal eye movements towards the contralateral side by stimulation of the posterior parts of area 7 and the lateral parts of area 5 (Fig. 39).

As these results indicate, electrical stimulation of area 5 produces complex movements of the limbs and trunk and occasionally the eyes, stimulation of area 7a various kinds of eye movements, stimulation of area 7b complex movements of the hand.

B. Man

In humans stimulation of the posterior parietal cortex rarely produces any responses (Penfield and Rasmussen 1950). However, in some older studies responses have been described. Bartholow (1874) produced contractions of the orbicularis oculi muscles and dilation of the pupils by faradic stimulation of the posterior parietal cortex of a patient. Foerster (1931, 1936a,b,c) made more systematic observations on responses from various cortical regions of non-anaesthetized humans (Fig. 40). He reported that strong stimulation of area 5b (PE of von Economo) elicited turning of the eyes and head towards the opposite side followed by simultaneous

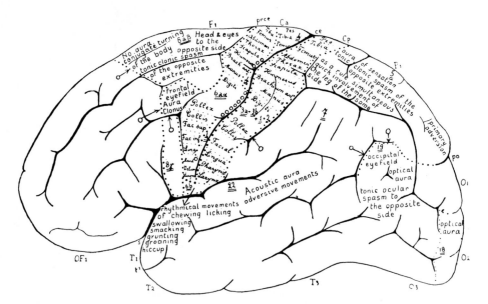

Fig. 40. Motor responses to electrical stimulation of the human cerebral cortex obtained by Foerster (1931)

flexion of the contralateral arm and leg and sometimes the ipsilateral leg when the current spread to the adjacent area 5a. An epileptic focus in the same region produced similar movements and paraesthesias and pain in the opposite side of the body. An epileptic focus inside the intraparietal sulcus produced a vestibular aura. Occasionally arm and hand movements were produced by stimulation of area 7b (PF of von Economo), but usually no responses were elicited from the inferior parietal lobule. Sensory experiences were never described after stimulation of the inferior parietal lobule (Foerster 1936c, Penfield and Rasmussen 1950).

VII. Neuronal Activity in Area 5

The function of neurones in area 5 of monkeys is related to the somatic sense and somatomotor behaviour. The somatosensory activity observed here is one degree more complex than that observed in posterior SI and described in Chapters III.B and III.C. In area 5 the sensory properties of neurones relate mainly to joint positions. The neurones activated from the skin are directionally selective to the movement on the skin or have other complex properties. All major studies published on area 5 neurones of monkeys (Duffy and Burchfiel 1971, Sakata et al. 1973, Mountcastle et al. 1975, Mackay et al. 1978) agree that neurones in area 5 are at a higher level in the chain of neural processing than those observed in SI. Duffy and Burchfiel (1971) and Sakata et al. (1973) emphasized the hierarchical nature of the sensory projections of these neurones, and Mountcastle et al. (1975) described their motor properties which they related to a "command function".

A. Sensory Properties

Duffy and Burchfiel (1971) recorded in paralyzed monkeys and observed that 63% of the neurones in area 5 were related to position sense only, 12% to touch only, and 12% to both touch and position sense. Most of the neurones related to position sense were activated optimally by simulataneous stimulation of several joints or joints and cutaneous receptive fields. *Natural combinations of positions in several joints constituted optimal stimulation* for many neurones. For instance, flexion of the hip joint could be an effective stimulus only if the knee was maintained in the extended position during the movement. Inhibitory interaction was also observed; e.g., flexion of the ipsilateral hip could prevent the excitatory response to flexion of the contralateral hip. Furthermore, opposite movements of two limbs could also act as best stimulation, for instance flexion on one side and extension on the other. For some neurones alternating movements such as produced by walking was the best stimulus.

Sakata et al. (1973) studied area 5 more extensively in non-anaesthetized, paralyzed monkeys. Of the 245 neurones 43% were related to joints,
15% to the skin, and 36% to both skin and joints. Most of the joint neurones were activated during combined rotation or positioning of several
joints. Both contralateral, ipsilateral and bilateral cutaneous receptive
fields were observed. Directional selectivity was characteristic for most of
the cutaneous neurones. Figure 41D presents an example of a directional
skin neurone in area 5 from the study of Sakata et al.

Since it is difficult to describe verbally the stimulus combinations
effective in sensory activation of area 5 neurones these stimuli are presented in Fig. 41 redrawn from the studies of Sakata (1975), and Sakata et al.
(1973). Optimal stimulus patterns for joint combination neurones are
illustrated in Fig. 41 A-C. The best stimulation for neurone A was a

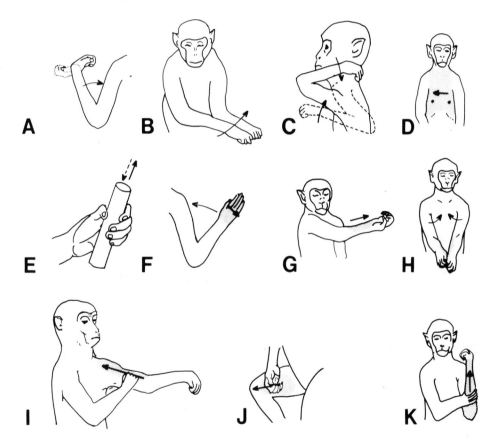

Fig. 41 A-K. Figurines indicating the most effective sensory stimuli for neurones studied
by Sakata et al. (1973, 1975) in area 5 of paralyzed awake monkeys. Each activation pattern is described in the text. The documents of impulse discharges in response to each
stimulus pattern can be found in the original papers. (Hyvärinen 1982)

combination of the flexion of the wrist and flexion of the elbow, whereas either flexion alone was ineffective when the other joint was extended. Neurone B illustrates bilateral interaction of joint stimuli; the best stimulus was parallel rotation of both arms from right to left. Neurone C was activated by elbow flexion when the shoulder was kept anteriorly elevated; it did not respond to the flexion of the elbow when the shoulder was posteriorly elevated (the position marked with dotted line).

Figure 41E-K illustrates the stimuli for neurones activated optimally by a combination of joint and skin stimulation. Neurone E gave the best response to a preaxially directed movement of a cylindrical object when the fingers were flexed against the object. The best response of neurone F was triggered by the simultaneous occurrence of flexion of the elbow and rubbing of the hand dorsum in the radial to ulnar direction. Such a stimulus combination occurs naturally when the hand dorsum touches something during active flexion of tbe elbow. The best response of neurone G was elicited by distally directed cutaneous stimulation when the shoulder joint was adducted; the same cutaneous stimulus was ineffective during abduction of the shoulder joint.

Sakata et al. (1973) called some neurones "matching neurones" because they were maximally activated by bringing two body parts into contact with each other. Neurone H was activated by adduction of either shoulder, but the best response was obtained when the palms or forearms were rubbed against each other while the shoulders were adducted. Neurone I illustrates a complex stimulus pattern consisting of flexion of the right elbow when the left shoulder was in the position indicated and the right arm was rubbing against the skin of the left shoulder in the direction from distal to proximal. Neurone J, on the other hand, was optimally activated when the dorsum of the right hand rubbed against the right thigh in the distal direction. The best stimulus for neurone K was rubbing the left forearm flexed at the elbow with the right hand in the direction from proximal to distal.

The complex stimulus patterns needed for the effective activation of some neurones suggest that the sensory properties of these neurones arise through convergence of inputs from a neural network probably partially located inside area 5 and partially in other somatosensory areas. The most direct way of explaining these properties is a hierarchical projection from neurones with less complex properties. The purpose of the neurones with such complex properties could be to signal automatically specific posture and movement patterns and thus aid the sensory guidance of purposeful motor acts.

Mountcastle et al. (1975) also observed the sensory properties of 977 area 5 neurones that they studied. They found that the majority of these neurones (64%) were activated by passive joint rotation and that passive

stimulation of muscles and other deep tissues activated 12% of the neurones. Another 12% were activated by cutaneous stimulation. Optimal
activation by simultaneous stimulation of several joints was less common
than in the studies of paralyzed monkeys. However, this result was probably due, at least partly, to the fact that behaving monkeys seldom tolerate
extensive passive movements in several joints for the long periods required
for satisfactory documentation. An important observation made by Mount
castle et al. (1975) was that area 5 neurones were much more active during
active movement than during passive joint stimulation. Furthermore, changes
in the level of alertness of the animal had a strong effect on area 5 neurones
which became undrivable during periods of drowsiness or sleep with return
of responses when the animal became alert again.

   Neurones activated by visual stimuli are rare in area 5. Mountcastle et
al. (1975) found in area 5 six visually activated neurones which constituted
only 0.6% of the neurones studied. Sakata (1975) described a few neurones
that could be activated by somatosensory and visual stimuli. The visual
responses of these cells were unstable whereas the somaesthetic ones were
stable. Interestingly, responses to both modalities exhibited similar directional preferences.

B. Motor Properties

   In the active, non-anaesthetized monkey Mountcastle et al. (1975)
found two new types of *neurones related to arm projection and manipulation*. These neurones comprised 9% of those studied in area 5, but in area 7
manual reaching, tracking and manipulation neurones, as described by Hyvärinen and Poranen (1974), comprised about one third of the neurones
studied. According to Mountcastle et al. (1975) the arm projection neurones
did not respond to external sensory stimulation but discharged at high rates
when the animal projected its arm towards a target in the immediate
extrapersonal space. The discharge of these neurones was not related to the
details of the movement but to the act of arm projection as a whole. The
hand manipulation neurones were not activated by passive stimulation
of the skin and joints but discharged at high rates during manual exploration performed for obtaining food or other objects.

   Mountcastle et al. (1975) trained monkeys to reach out and press a
panel with a target light to obtain a juice reward. During the performance
of this task the activity of the arm projection neurones started to increase
when the monkey detected the target light, reached a peak when the arm
was moving towards the target and declined close to zero just before the

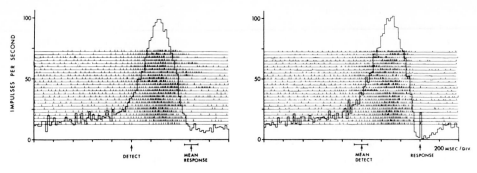

Fig. 42. Activity of an arm projection neurone in area 5 in an experiment of Mount-castle et al. (1975). Replicas and histograms of the responses during conditioned reaching movements towards a lighted push-button that delivered juice to the monkey. On the *left* the responses are aligned by superimposing the moments of detection of the light, on the *right* by superimposing the moments of reaching the response button. The response started before release of the detect key, reached its peak when the arm was on the way to the target, and declined close to zero shortly before contact with the target

hand contacted the target (Fig. 42). These neurones remained silent when the monkey was not performing arm projection movements motivated by rewards. The activity pattern of these neurones was independent of the particular spatial trajectory of the arm movement towards the target.

Three possible interpretations of the arm-projection and hand manipulation neurones were discussed by Mountcastle et al. (1975). They rejected the possibility that these neurones were the same ones that were activated by passive joint and other stimuli, because of the lack of responses to such stimuli. Because the responses varied greatly and depended on the motivation of the animal, the authors favoured the hypothesis that these neurones act as a command apparatus for manual exploration of the immediately surrounding extrapersonal space. The command function concept implies a decision-making unit whose activity arises from a combination of sensory and motivational inputs and leads to motor action (Fig. 43). The concept of "command neurones" was criticized and discussed extensively by Kupfermann and Weiss (1978) and in the commentaries of their paper. However, some of the functional features of arm projection and hand manipulation neurones could fit in a "command system" as suggested by Mountcastle et al. (1975).

According to the third interpretation the activity of these neurones represented a "corollary discharge" or "efference copy" of the motor command discharge from the motor regions of cortex. Various interpretations of the functional role of these neurones will be dealt with in Chapter XII.

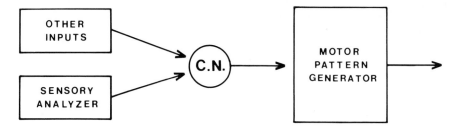

Fig. 43. A diagram illustrating the concept of a command neurone *(C.N.)*. (Kupfermann and Weiss 1978)

C. Sensorimotor Interaction in Area 5

It is tempting to think that the sensory and motor aspects of the activity of area 5 neurones are different sides of the same function. The sensory properties of these neurones are easier to investigate in the paralyzed animal, but in the alert, behaving animal the same neural circuits are driven to action during active movements. Area 5 seems to have an important role in *somatosensory exploration, reaching for and manipulation of objects.* Sensory feedback is necessary for a coordinated performance of such acts. Sensory guidance of purposeful acts may be facilitated by a sensory neural network where the signal processing is automatized, for instance on the basis of hierarchical convergence. For successful reaching for different targets the positions of several joints have to be coordinated, but when performing such acts we are unaware of the details of the movements in various joints. Only the location of the goal is important. The same principle seems to apply to many area 5 neurones. The cellular machinery here appears to carry out a synthesis of the individual elements of the movement to form a purposeful entity which secures that the end is attained. The neural circuits connecting the limbic and hypothalamic structures with the posterior parietal lobe (Chap. IV) could mediate motivational and intentional influences to this region. Inputs related to motivation or motor corollary discharge could act as gates allowing sensory input from the joints and elsewhere to reach area 5 neurones. Such a model could explain the enhancement of the activity of these neurones during motor acts.

VIII. Neuronal Activity in Area 7

A. Visual and Oculomotor Mechanisms

In the 1960's the posterior parietal cortex was considered as a somatic associative area. This opinion was based on what was known of anatomical connections, ablation studies and recordings from area 5. However, when cellular recordings from area 7 of behaving monkeys were first started in Helsinki in 1970, we found to our surprise that many neurones were activated by visual stimuli. The preliminary findings were described by Hyvärinen and Poranen (1974) as follows: "Cutaneous activation from the contralateral side of the face was an effective stimulus for some area 7 cells, but they sometimes discharged action potentials before their peripheral receptive fields had been touched. This 'anticipatory activation' was not due to weak air streams set in motion by the approaching stimuli as shown by the fact that the 'anticipatory activation' was abolished when the animal was prevented from seeing the approaching stim-

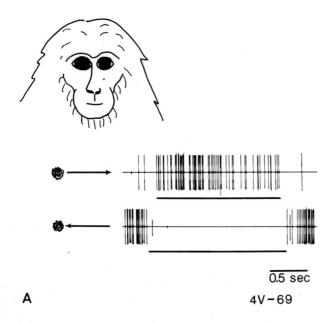

A   4V – 69

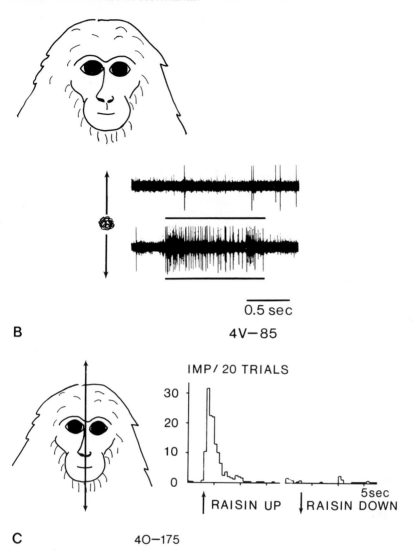

0.5 sec

B   4V—85

IMP/ 20 TRIALS

↑ RAISIN UP   ↓ RAISIN DOWN

C   40—175

Fig. 44 A-C. Neurones whose activity was related to eye movements. **A** Activity (electronically shaped pulses) of a left area 7 cell recorded when the monkey was following with its eyes a raisin moving back and forth to the left and to the right. During movement to the left the cell discharged, and the discharge ceased abruptly when the direction of the movement was reversed. **B** Cellular discharges recorded from another left area 7 cell activated during visual pursuit of a raisin downward. **C** Histograms of cellular responses of a right area 7 cell during visual pursuit of a raisin moving rapidly up and down in front of the monkey. (Hyvärinen and Poranen 1974)

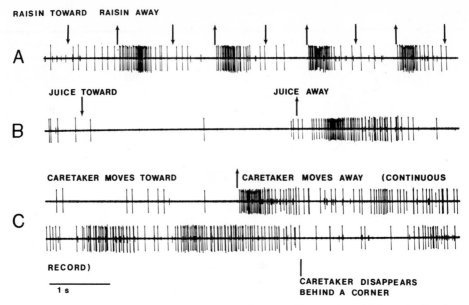

Fig. 45 A-C. Visual responses of a cell in area 7 responding to the withdrawal of various targets from the monkey. **A** Responses to the withdrawal of a raisin from the monkey. **B** Responses to the withdrawal of a spoon filled with orange juice. **C** Activity during movements of the monkey's caretaker. (Hyvärinen 1981a)

ulus. As long as vision was blocked only tactile stimuli to the cutaneous receptive field discharged such cortical cells. Thus we concluded that visual activation was capable of driving the cells in area 7. Light stimuli flashing on or off at a distance did not discharge these cells, nor did contoured stimuli approaching from the other side of the face (contralateral to the cutaneous receptive field). It therefore seemed that the visual stimulus had to emerge in a location close to the cutaneous receptive field of the cell."

These and other preliminary observations of the activity of area 7 neurones were presented in August 1971 in a Satellite Symposium of the Munich Congress of Physiological Sciences on the Somatosensory System in Ulm, Germany (Hyvärinen et al. 1975), in 1972 in The Neurosciences: Third Study Program, Boulder, Colorado (Hyvärinen et al. 1974) and in 1973 at the Scandinavian Physiological Congress in Bergen, Norway (Hyvärinen 1973).

In our first full report on area 7 neurones (Hyvärinen and Poranen 1974) we also reported neurones active during eye movements in specific directions (Fig. 44). Many of these neurones gave a response to passive visual stimulation but many were also strongly activated during eye movements. Therefore, we called these neurones "looking" neurones in analogy with the reaching and manipulation neurones related to somatic functions. In these neurones a visual sensory mechanism was assumed to operate during active visual searching movements. Figure 45 presents a neurone, belonging to this group, whose activity was related to the distance of the object under observation. This neurone responded to visual stimuli moving away from the monkey. The monkey was visually fixating the moving objects; thus the discharge correlated with divergence eye movements. However, during movement of the monkey's caretaker the firing rate of this neurone stayed increased until the caretaker disappeared from the room at a distance of 5 to 6 m. At this distance the divergence eye movements were probably not the sole determinant of the discharge; the diminution of the size of the visual target may also have influenced the activity of this neurone.

Mountcastle et al. (1975) confirmed the visual activation in area 7 neurones which thus differed from the neurones in area 5, but they suggested an eye movement command function for these neurones. Later studies have emphasized both the visual sensory and oculomotor functions of area 7 neurones; now most authors appear to agree that the role of these neurones relates both to sensory and to motor functions. In this sense they appear truly "associative".

Mountcastle et al. (1975) trained monkeys to fixate and pursue light targets for liquid reward and recorded the eye position using electro-oculography. They described three neurone types related to vision and eye movements. These were called "visual fixation neurones" discharging when the animal fixated interesting objects, "visual tracking neurones" active during smooth pursuit of interesting objects within the arm's reach and "visual space neurones" active when visual stimuli were presented lateral to the maximal deviation of gaze. Yin and Mountcastle (1977) called the latter group "light-sensitive". Lynch et al. (1977) described yet another group called "saccade neurones". However, th relationship of these neurones to fixation, tracking and saccades was seriously questioned by Robinson et al. (1978) who explained the properties of these neurones on the basis of their visual receptive fields.

It now appears that most of the features of area 7 neurones are best explained on the basis of the sensory visual properties of these neurones. However, some results, e.g., discharges related to eye movements in the dark may best be explained in connection with eye movements. As suggested by Motter and Mountcastle (1981) there may exist chains of neurones leading gradually from purely sensory to motor properties.

1. Visual Fixation Neurones

The discharge rate of these neurones increased rapidly when the monkey fixated objects. One third of the neurones studied in area 7 by Mountcastle et al. (1975) belonged to this group. In other studies their proportion of area 7 neurones has ranged from 11% (Sakata et al. 1980) to 57% (Lynch et al. 1977). According to Mountcastle et al. (1975), these neurones were not driven by passive visual stimulation, and their activity was suppressed during saccades.

Lynch et al. (1977) defined gaze fields for visual fixation neurones as that zone in space in which fixation of a target light was associated with increased activity of the neurone. Their fixation neurones were of two types, those with limited gaze fields and those with full gaze fields. The gaze fields could be limited to one half or quadrant of the visual field, usually on the contralateral side. Neurones with full gaze fields were active during fixation in any direction.

The interpretation of Robinson et al. (1978) was quite different. They observed that area 7 neurones had large visual receptive fields that usually covered the contralateral visual field and could include the fovea. Best responses were obtained using large, bright stimuli; the response latencies were about 100 to 120 ms. According to Robinson et al. (1978), these neurones responded to sensory stimuli in the absence of eye movement but did not discharge in association with movement in the absence of a visual stimulus. These authors explained the visual fixation neurones with limited gaze fields as sensory neurones which became active when the direction of gaze caused a visual contour of the environment to excite their receptive fields. This explanation was supported by their finding that lowering of the illumination reduced the firing of limited gaze field neurones. Robinson et al. (1978) also regarded the fixation neurones with full gaze fields as sensory, having receptive fields in the fovea and discharging tonically in response to foveal stimuli. They could demonstrate sensory responses in these neurones by varying the intensity of the fixation light.

Sakata et al. (1980) studied 125 fixation neurones in the posterior part of area 7a in the anterior lip of the superior temporal sulcus. This region was not sampled in the studies of Mountcastle et al. (1975), Lynch et al. (1977), or Robinson et al. (1978). The fixation neurones comprised 11% of the sample of 1139 neurones recorded by Sakata et al. (1980). The discharge rates of these neurones proved to be monotonic functions of the angle of gaze. Thus no specified gaze fields were detected for these neurones. In addition, many fixation neurones were selective for the depth of fixation or distance from the animal; some were selective for both the radial direction and distance. The discharge rate of one half of the fixation neurones decreased in the dark, suggesting that sensory visual stimuli from the environment influenced their activity. The other half was equally active in the dark as in the light and their discharge rates also correlated closely with eye position in the dark. Their activity could not be explained on the basis of visual signals. Sakata et al. (1980) therefore suggested that these neurones received *extraretinal input signaling the direction of gaze.* They furthermore suggested that these neurones, by integrating the retinal and extraretinal signals, aid the discrimination of three-dimensional positions of objects in the visual space and the control of the direction and distance of fixation.

The results of Sakata et al. (1980) differ clearly from those of Lynch et al. (1977) and Robinson et al. (1978). Lynch et al. (1977) suggested limited or full gaze fields, whereas Sakata et al. found a monotonic relation between the discharge rate and the angle of gaze. This difference could probably be explained by the different locations of the recordings. However, in a recent study of the anterior part of area 7a, Motter and Mountcastle (1981) agreed with Sakata et al. (1980) on the gaze fields of fixation neurones. In the study of Motter and Mountcastle (1981) the fixation neurones comprised 16% of the cells studied; they were insensitive to passive visual stimulation. Another 4% of the neurones studied were both sensitive to visual stimulation and activated by fixation.

On the basis fo the above evidence area 7 appears to contain neurones related to visual fixation. However, they are less common than was suggested by Lynch et al. (1977), whereas in agreement with Robinson et al. (1978) sensory stimulation influences more neurones than Lynch et al. (1977) suggested. These neurones may have an extraretinal sensory input, and they may also receive a corollary discharge or efference copy from the structures that regulate eye movements as discussed by Sakata et al. (1980).

2. Visual Tracking Neurones

Mountcastle et al. (1975) observed neurones that were active during smooth pursuit of interesting objects within the arm's reach. The majority of these neurones were active during movements in one direction only, but all different directions were represented in different neurones. These neurones comprised 7% and 8% of the neurones in area 7a in the studies of Lynch et al. (1977) and Sakata et al. (1980), but the sample of Mountcastle et al. (1975) contained 20% of them and the sample of Motter and Mountcastle (1981) only 1%.

Robinson et al. (1978) regarded the tracking neurones as sensory neurones discharging tonically when the stimulus moved in the receptive field in a direction specific for the neurone. According to them these neurones responded equally well when the animal made a pursuit eye movement which moved a contour of the surroundings over the receptive field as when the animal fixated a stationary spot of light and a visual stimulus moved over the stationary receptive field. However, Lynch (1980b) pointed out that the tracking neurones often start to fire before the onset of the eye movement.

When studying the posterior part of area 7 Sakata et al. (1978) found 20 tracking neurones, 6 of which were excited during tracking of a target in a particular direction, and also by the movement of a visual frame around the target in the opposite direction while the target of fixation was kept stationary. This finding demonstrates the role of the visual background when defining the target of eye movements and reminds one of the illusion of induced movement. The target and the background differ from each other by moving differentially each as a coherent group of contours.

The tracking neurones also appear to receive both retinal and extraretinal signals. The tracking neurones studied by Sakata et al. (1978) were also activated in complete darkness when there was no moving surrounding image. In this situation the most likely source of input was the extraretinal signal of eye movements. These neurones responded thus both during the movement of the visual frame and during tracking in darkness, indicating that retinal and extraretinal signals converge upon them (Shibutani et al. 1982).

3. Saccade Neurones

Saccade neurones in area 7 were described by Lynch et al. (1977) as neurones active before and during visually evoked saccades but not before spontaneous saccades. Many of them were activated during saccades towards the contralateral side. Their discharge preceded the onset of the eye movement on the average by 73 ms. According to Lynch et al. (1977), these neurones comprised 17% of the neurones in area 7.

The interpretation of Robinson et al. (1978) was that saccade neurones respond to visual stimuli when these stimuli excite their receptive fields during eye movements. They observed that these neurones did not fire before saccadic eye movements made in the absence of the proper stimulus. The responses of half of these neurones to stimuli on the receptive fields were enhanced when those stimuli became targets of eye movements.

Sakata et al. (1980) did not mention saccade neurones, but in their new study Motter and Mountcastle (1981) confirmed their existence. However, only 4% of the neurones studied belonged to this group. Motter and Mountcastle (1981) also confirmed the finding of Robinson et al. (1978) that many neurones responsive to passive visual stimulation were also influenced by saccadic movements. Their responses to visual stimuli were enhanced or suppressed when those stimuli became targets for saccadic eye movements.

4. Visual Sensory Neurones

Visual sensory responses were described by Hyvärinen and Poranen (1974) in neurones also responding to cutaneous stimulation (Fig. 46). Mountcastle et al. (1975) identified a class of neurones which they called "visual space neurones", that were activated by visual stimuli presented lateral to the maximal deviation of gaze. These neurones were called "light-sensitive" by Yin and Mountcastle (1977) who found that 28% of the neurones in area 7 belonged to this class. These neurones had large peripheral visual receptive fields that did not include the fovea.

Robinson et al. (1978) claimed that most area 7 neurones had visual receptive fields and that all the relationships of these neurones to eye movements could be explained on the basis of their visual sensory properties. In other studies visual sensory neurones have comprised 30% to 40% of the neurones recorded in area 7. However, the representation of visual mechanisms is more common in the medial part of area 7 (Hyvärinen and Shelepin 1979, Hyvärinen 1981c). Thus the proportion of visual sensory neurones may also depend on the location of the sample in area 7. The high percentage of visual sensory neurones reported by Robinson et al. (1978) probably results from the automatic summation of the visual responses that was performed for all neurones studied, whereas others have classified neurones into different groups on the basis of qualitative observations prior to setting up computer-controlled sampling of the data. Weak visual responses may be recorded by automatic summation method also when the neurones are responding optimally to other stimuli or during motor performance.

It was observed in several studies in this laboratory that neurones which were directionally selective to moving visual stimuli were often also activated by cutaneous and joint stimuli to which they showed directional preferences, too (Fig. 47) (Hyvärinen and Poranen 1974, Leinonen et al. 1979, Leinonen and Nyman 1979, Leinonen 1980). Motter and Mountcastle (1981) have now studied directional sensitivity in the visual

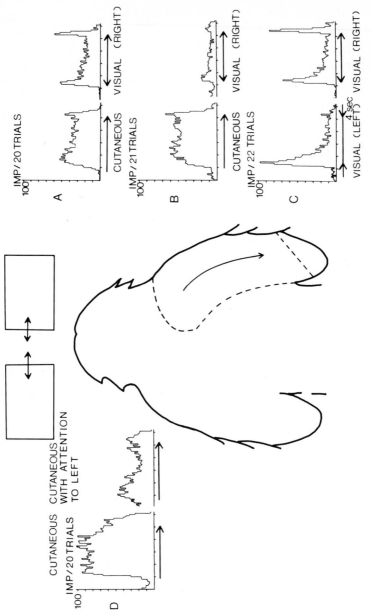

Fig. 46 A-D. Response histograms of a neurone in left area 7 that responded to visual and cutaneous stimuli. **A** Responses to alternating cutaneous stimuli moving distally over the right shoulder and to visual stimuli consisting of a white card brought rapidly into the visual field from the right (on-response), then kept stationary (sustained response), and rapidly withdrawn to the right (off-response). **B** Same as **A** but the view of the monkey almost totally blocked. **C** Responses to alternating presentations of white cards from the left and from the right. **D** Responses to cutaneous stimuli moving distally over the right shoulder. During every other cutaneous stimulus orange juice was offered from the left. Visual attention to the left caused a decrease in the cutaneous response. (Hyvärinen and Poranen 1974)

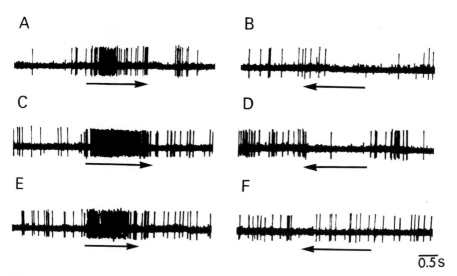

Fig. 47 A-F. Responses of a cell in area 7b to moving visual and cutaneous stimuli. In **A**, eyes covered, the monkey's chest was stroked with a finger from right to left and in **B** from left to right. In **C** and **D**, eyes open, the stimuli are the same as in **A** and **B**. In **E** a visual stimulus (the investigator's hand) moves from right to left at a distance of 15 cm from the chest and in **F** from left to right. (Leinonen et al. 1979)

neurones of area 7a using quantitative techniques with very interesting results. In that study visual sensory neurones comprised 31% of the neurones in area 7a. They had large, often bilateral receptive fields and discharged with 50 to 290 ms latencies to visual stimuli. Most of these neurones showed the so-called *"foveal sparing"* phenomenon which meant that stimuli centered on the fixation point evoked no responses. Most neurones were sensitive to moving visual stimuli and 90% were directionally selective. The preferred directions of movement were called "vectors". Most of the vectors were directed either towards or away from the fixation point. For the neurones with bilateral receptive fields the vectors of the two half-fields pointed in opposite directions arranged in a radial manner. This arrangement was called *"opponent vector orientation"*. The velocity sensitivity of these neurones to moving visual stimuli ranged from $10°$ to $800 °/s$, but within this range the responses to different velocities did not differ. This velocity range matches the angular velocity of visual stimuli during locomotion.

Moreover, responses to visual stimuli at identical retinotopic loci varied as a function of the angle of gaze. The effect of the angle of gaze was also present in the dark and was thus not mediated by the visual contours of the environment. Thus these neurones are likely to receive extraretinal sensory input or corollary motor discharge. Furthermore, visual fixation facilitated responses from extrafoveal receptive fields, whereas those from the fovea were inhibited. As Motter and Mountcastle (1981) suggest, these visual sensory mechanisms are well suited for the functions of the so-called *ambient vision* (Trevarthen 1968, Schneider 1969), which serves head and body orientations, postural adjustment and locomotory displacements that change the relationships of the body and its parts in reference to the surrounding objects during visual fixation and foveal work. The facilitation of extrafoveal responses by fixation suggests that area 7 monitors the peripheral visual field during foveal work to detect movements there. The detection of peripheral movement may then lead to a shift in the direction of gaze towards the moving stimulus.

Kawano and Sasaki (1981) have recently showed that optokinetic stimulation, i.e., rotation of visual surround, is an effective stimulus for many visual neurones in area 7. These neurones were not observed to receive vestibular input. However, since the receptive fields were not delineated and smaller stimuli not used, it is possible that for some of these neurones stimuli of lesser width than the total surround might also have been effective.

B. Somatic Mechanisms

In the lateral parts of area 7, particularly in area 7b, there are neurones that respond to cutaneous or to kinaesthetic stimuli or to these and visual stimuli. In the most lateral part of area 7b in the region of the parietal operculum next to the second somatosensory area a purely somaesthetic part of area 7b was described (Robinson and Burton 1980b, c). In the posterior part of area 7a a kinaesthetic projection zone was located (Hyvärinen 1981c). Neurones related to somatic movements, arm projection, hand manipulation, and activity of the lips are also distributed in various parts of area 7.

1. Cutaneous Responses

Cutaneous receptive fields for neurones that were activated by visual stimuli were described by Hyvärinen and Poranen (1974) (Fig. 46), and a few such neurones were also found by Mountcastle et al. (1975). In a study of area 7b Leinonen et al. (1979) described neurones activated by cutaneous stimuli which had large receptive fields covering typically entire

Fig. 48. A Cutaneous receptive field of a directionally selective cell in area 7b. *Arrows* indicate the direction of movement of effective stimuli. **B** The mean firing rates and standard deviations of the cell as a function of the speed of the stimulus movement. The stimulus, light stroking with a finger along the skin from the elbow to the fingertips, was repeated five times at each speed. (Leinonen et al. 1979)

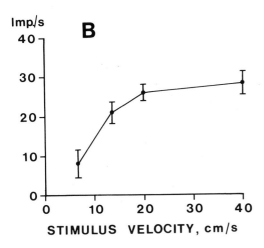

parts of the body such as the arms and hands or feet and legs. Figure 48 illustrates properties of a neurone that responded to cutaneous stimuli on the body above the waist level and directionally selective for movement along the skin towards the nose or fingers. Movement velocity of over 20 cm/s evoked maximal responses from this neurone. Several neurones were directionally selective for movement over the skin. The receptive fields are indicated in Fig. 49. In the posterior part of area 7 Leinonen (1980) found receptive fields mainly on the arms and hands (Fig. 50), whereas in the lateral part most receptive fields were on the face (Fig. 51) (Leinonen and Nyman 1979). These neurones differed from those in SI in that their responses quickly diminished on repetitive

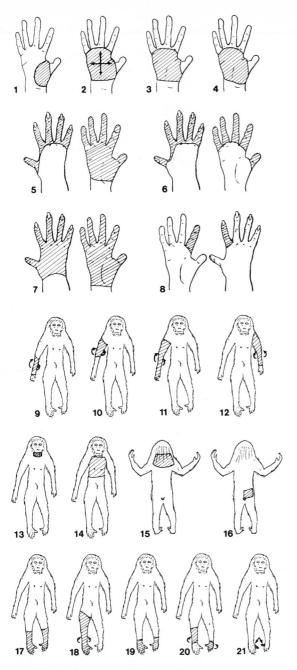

Fig. 49. Receptive fields (*hatched*) of 28 area 7b neurones that responded only to cutaneous stimulation. For simplicity, the right side of the figurines represents the side contralateral to recording. *Short arrows* near the receptive fields indicate that the receptive field covers also the backside of the limb or body. *Arrows* on the receptive fields (cells no. 22-28) indicate the direction of movement of effective stimuli. (Leinonen et al. 1979)

A.

1–11   13–18

B.

1–5

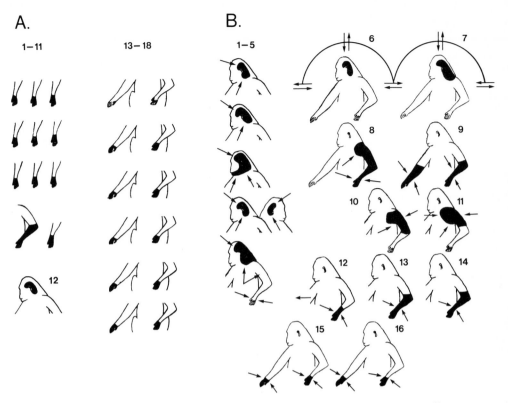

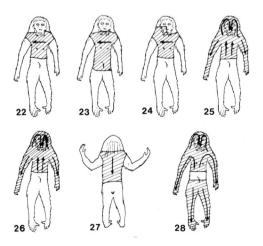

Fig. 50 A, B. Receptive fields of neurones in posterior area 7. **A** Neurones responding only to touching or compressing of the skin. **B** Neurones that responded both to touching or compressing of the skin and to visual stimuli. *Arrows* indicate effective directions of movement of the visual stimuli. Cells Nr 6 and 7 responded to visual stimuli moving in the periphery of the visual field. (Leinonen 1980)

22   23   24   25

26   27   28

Fig. 49 (continued)

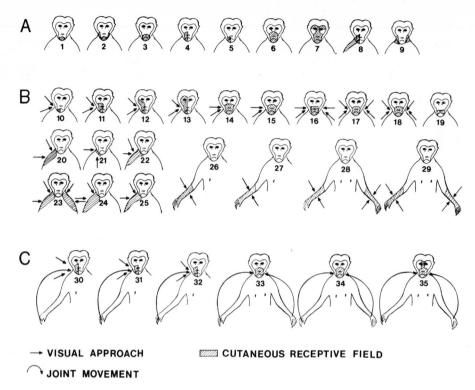

→ **VISUAL APPROACH**   ▨ **CUTANEOUS RECEPTIVE FIELD**

↷ **JOINT MOVEMENT**

Fig. 51 A-C. Receptive fields of neurones in the lateral part of area 7b, the associative face region. **A** Receptive fields of neurones that responded only to cutaneous stimulation. **B** Receptive fields of neurones responding both to touching the skin and visual stimuli approaching the cutaneous receptive field. One neurone (No. 19) was active during convergent visual fixation. **C** Receptive fields of neurones that responded to both cutaneous and proprioceptive stimulation. Three of them responded to approaching visual stimuli and three were active during convergent fixation (Nos. 33, 34, 35). (Leinonen and Nyman 1979)

stimulation. Their receptive fields were much larger than those found in SI, often bilateral, and they responded better to stimuli with fairly large surface area. Robinson and Burton (1980b) found that the neurones in 7b also differed from those in SII; their receptive fields were more variable, several different body regions were represented, and the receptive fields recorded along a penetration did not progress in an orderly sequence. Moreover, the area of the receptive fields often diminished during sleep or anaesthesia. Noxious stimulation was the adequate stimulus for 7% of the neurones recorded in the lateral part of area 7b. Attention towards the stimulus had a clear effect on the responses of the skin neurones in this part of area 7 (Robinson and Burton 1980c).

2. Kinaesthetic Responses

Neurones activated by joint rotation or palpation of muscle bellies were also recorded in area 7. Such neurones have comprised from 15% to 25% of those recorded in the lateral part of area 7 (Leinonen et al. 1979; Leinonen and Nyman 1979; Leinonen 1980). In the posterior part of area 7 the muscles and joints evoking responses were mainly on the upper extremities (Leinonen 1980), whereas the muscles the palpation of which activated neurones in the lateral part were located in addition around the mouth and on the neck and shoulders (Leinonen and Nyman 1979). Figure 52 illustrates three consecutive responses to the squeezing of a muscle belly recorded in a neurone in the posterior part of area 7. The figure shows that the contralateral response was much stronger than the ipsilateral one and that these responses habituated quickly.

In a mapping study of responses in various parts of area 7 I found that the responses to stimulation of joints and muscles were concentrated in the posterior part of area 7a (Fig. 74A) (Hyvärinen 1981c).

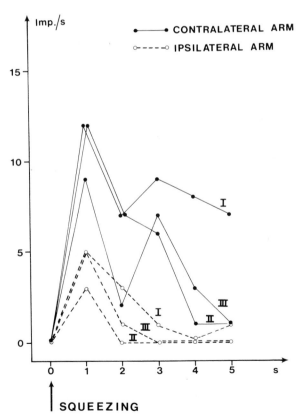

Fig. 52. Firing rate of a neurone in posterior area 7 activated by palpation of the ulnar muscle belly. Continuous squeezing of the ulnar muscle belly was performed three times successively on the contralateral arm (*I, II, III, solid lines*) and three times successively on the ipsilateral arm (*I, II, III, dotted lines*). Fairly rapid adaptation of the response is seen. (Leinonen 1980)

3. Activity Related to Somatic Movements

Discharges related to manual *reaching, tracking and manipulation* were described by Hyvärinen and Poranen (1974). An example of a neurone activated during manual tracking downwards is given in Fig. 53. This neurone was activated by hand tracking movements downwards but not upwards, and the activation was independent of the hand used. When the hands were fixed and the monkey followed the moving raisin with its eyes, eye movements up and down alone were not sufficient to activate this cell, although covering the eyes seemed to reduce the responsiveness of the cell during somaesthetically guided reaching. Most neurones related to manual reaching were activated better when the reaching was performed using the arm contralateral to the recording.

In the lateral part of area 7b some cells were activated during reaching with the lips (Fig. 54) (Leinonen and Nyman 1979). In this region biting and elbow flexion also activated neurones. Neurones related to manipulation with fingers are different from those related to reaching. These neurones probably get part of their activation from the skin during active

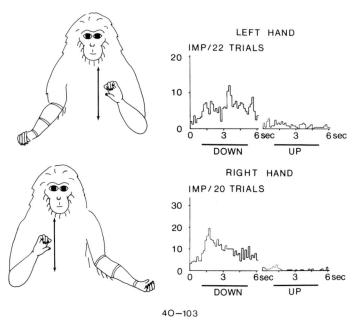

Fig. 53. Histograms of responses in a right area 7 neurone during reaching for a raisin which is being moved up and down. In the *upper figure* the monkey reached for the raisin with the left hand (right arm fixed); in the *lower figure* it reached with the right hand (left arm fixed). The cellular activation was good when the monkey was pursuing the target with either hand downwards but not when pursuing upwards. Eye movements alone did not activate this cell. (Hyvärinen and Poranen 1974)

A   B

0.5 s

Fig. 54 A, B. Discharges of a neurone in the lateral part of area 7b, associative face region. The cell was active during reaching movements performed with the lips. **A** The desired object was on the contralateral side of the mouth (lips did not touch the object). **B** The desired object was on the ipsilateral side. (Leinonen and Nyman 1979)

A   B

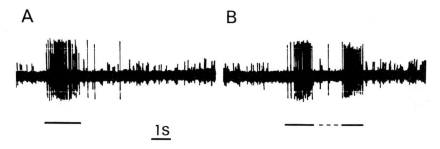

1s

Fig. 55 A, B. Activity of a neurone in area 7b; the cell discharged during manipulation with the fingers. In the figure, activity starts when the animal touches a raisin held by the investigator, lasts as long as the manipulation continues, and stops immediately when the monkey gets hold of the raisin, i.e., during the isometric contraction of the finger flexors there was no activity. **A** Manipulation with the contralateral fingers (———). **B** Manipulation with the ipsilateral fingers, (———) marks a pause in the manipulation. (Leinonen et al. 1979)

manipulation, although it is not possible to delineate a receptive field for them by using passive stimuli. An example is given in Fig. 55. This neurone was activated during manipulation with either hand.

The incidence of the neurones related to somatic movements has varied between 8% and 36% in different studies of area 7. Their percentage has been greater in the lateral parts of area 7 than in the medial parts. Neurones related to hand manipulation are more common in the lateral parts of area 7.

C. Convergence of Somatic and Visual Functions

In the lateral part of area 7 many neurones are activated by both somatic and visual stimuli. The incidence of such neurones has ranged between 9% and 33% of those studied in the lateral part of area 7 (Hyväri-nen and Poranen 1974, Leinonen et al. 1979, Leinonen and Nyman 1979, Leinonen 1980, Robinson and Burton 1980c). Hyvärinen and Poranen (1974) described neurones activated by visual and cutaneous stimuli (Fig. 46), and others activated during both looking and reaching. Figure 56 illustrates the function of a neurone related to both visual and somatic activation in the direction furthest left, close to the left ear. While the monkey's arms were fixed, visual stimuli presented from this direction in the left peripheral visual field and which evoked the monkey's interest elicited a clear response; a similar stimulus from the right elicited no response, or possibly a slight inhibition. When its left hand was released and the monkey tried to grasp the raisin, a response was seen provided that the target was to the left in the region of the shoulder and the ear. When the target was in any other direction (for example Fig. 56, lower B) no response was obtained. The response was equally good when its eyes were

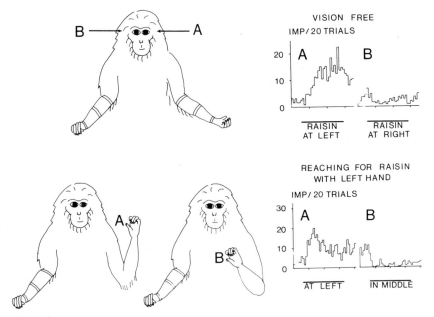

Fig. 56. Response histograms of a neurone in right area 7. This cell was activated by visual targets on the left side (*top A*) of the face and during manual reaching movements towards that region *(bottom A)*, but not elsewhere *(bottom B)*. (Hyvärinen and Poranen 1974)

covered and the monkey reached with its hand for a target in the activating region.

Findings of this type led us in 1974 to hypothesize that these neurones were specific to the direction of the aimed movement. In this sense they would have been analogous to the visual fixation neurones with limited gaze fields discovered later by Lynch et al. (1977). However, such directional specificity for the reaching neurones was denied by Mountcastle et al. (1975), and later studies indicate that limited gaze fields may result from visual sensory mechanisms. Whether the neurones in area 7 have specificity for the direction of reach or fixation is thus an open question at the moment and awaits further experimental work.

Several interesting examples of visual and somatic convergence were described by Leinonen et al. (1979). Figure 47 illustrates an example of a neurone responding to cutaneous and visual stimuli moving from right to left. Stimuli moving in the opposite direction did not activate this neurone. Thus this neurone exhibited a similar directional preference for stimuli of two modalities and could be said to generalize the direction of movement in these modalities.

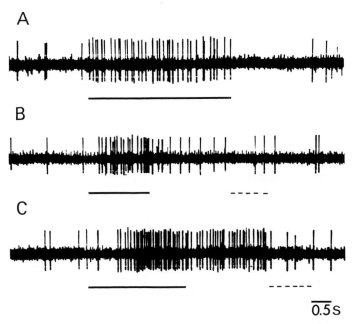

Fig. 57 A-C. Discharges of a neurone in the lateral part of area 7b, associative face area. A Eyes closed; responses to touching of the chin (——). B Eyes closed; responses when the monkey's hand was passively brought towards the chin (——), kept stationary near it (   ) and moved away (– – –). C Eyes open; responses when a visual stimulus approached the monkey's chin (——), was stationary near it (   ) and moved away (– – –). (Leinonen and Nyman 1979)

Another example of visual and somatic convergence with a common property linking the various stimuli together is shown in Fig. 57. This neurone recorded in the lateral face region in area 7b responded to touching of the snout, to visual stimuli approaching the mouth or remaining stationary close to the mouth, and to passive elbow flexion bringing the monkey's hand towards the mouth. The common property for all these stimuli was thus the direction of the movement towards the mouth around which was also the cutaneous receptive field. The cutaneous receptive fields and effective directions of movement are presented in Fig. 51C for several neurones of this type found in the lateral part of area 7b (Leinonen and Nyman 1979).

Yet another interesting type of visual and somatic convergence is presented in Fig. 58. In this neurone, as well as in several others, visual

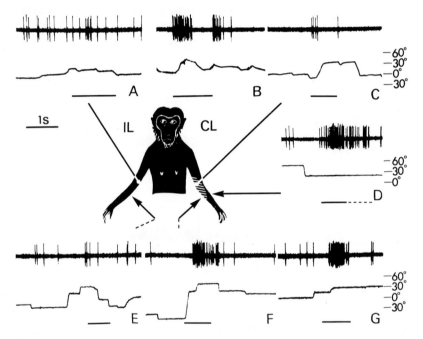

Fig. 58 A-G. Activity of a neurone in area 7b responding to touching of the contralateral arm and to visual stimuli approaching or staying near the contralateral arm. The trace below the cellular activity shows the horizontal electro-oculogram; positive deflections indicate eye movements in contralateral direction. The monkey's eyes are covered in **A,B,** and **C. A** Touching of the ipsilateral arm (——). **B** Touching of the contralateral arm (——). **C** Fifth successive touching of the receptive field (*hatched*) indicating rapid habituation of the response. **D** Experimenter's hand approaches the contralateral arm (——) from the contralateral side and stays above it (– – –). **E** Experimenter's hand approaches the ipsilateral arm (——) from the midline and is then moved immediately away. **F** Experimenter's hand approaches from the midline the contralateral arm (——). **G** Same as **F** repeated. (Leinonen et al. 1979)

stimuli approaching the cutaneous receptive field evoked responses. This neurone had a cutaneous receptive field on the contralateral arm. A light touching of the receptive field resulted in tonic discharge if the monkey could see the touching object, and in an on-off response when the monkey's eyes were covered (Fig. 58B, C). The response diminished on repetitive touching (Fig. 58B compared to 58C). The neurone also responded when the experimenter's hand (or an object in the experimenter's hand) approached the contralateral arm from any direction (Fig. 58D, F, G) or stayed near it (Fig. 58D). The monkey was watching these stimuli, but the examination of the horizontal (in the figure) and vertical eye movements revealed that the direction of the gaze per se was not the determinant factor of the responses. Visual stimuli approaching the ipsilateral (right) arm were ineffective (Fig. 58E) as well as visual stimuli approaching other body parts. Other examples of visual stimuli that were effective when approaching various body parts are given in Fig. 50B. In most such neurones the effective visual movement was towards or in close proximity to the body part where the cutaneous receptive field was.

Such observations indicate that are 7 participates in *intersensory associative analysis of directions of movement in the visual environment in reference to body parts*. This type of sensory information processing is needed for the motor guidance of movements of body parts, particularly the hand, towards targets.

D. Behavioural Mechanisms

Hyvärinen and Poranen (1974) noted that a prerequisite for activity in area 7 neurones was that the animal was *interested* in the sensory targets presented. Mountcastle et al. (1975) made similar observations and were of the opinion that the responses of these neurones were conditional on the stimulus objects and depended on the motivational set of the animal. Rolls et al. (1979) studied quantitatively the significance of the presumed motivational factors for the activity of 73 visual fixation neurones in area 7. They presented various rewarding, aversive and neutral visual stimuli through a shutter, but did not find differences in the responses to these stimuli. Moreover, feeding the monkey to satiation did not abolish responses to the presentation of food, as happened in hypothalamic neurones also studied by Rolls. These authors concluded that the activation of the fixation neurones was sensory in nature and responses were always obtained, provided the monkey was adequately looking at the stimulus. Rolls et al. (1980) also pointed out the danger of a circular argument if the

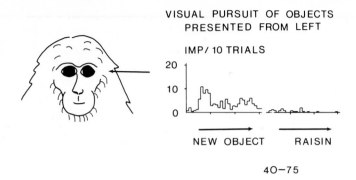

Fig. 59. Response histograms of a neurone in the right area 7 activated during looking to the left. Visual pursuit of a familiar object such as a raisin presented from the left did not activate the cell, but the same pursuit movement did activate the cell when a new object was presented. (Hyvärinen and Poranen 1974)

visual fixation itself was taken as the indication of the monkey's interest in the stimulus object.

However, a number of qualitative observations made by different authors suggest that many area 7 neurones are influenced by motivational and other behavioural variables. For instance Robinson and Burton (1980c) showed that the extent of the cutaneous receptive fields of the neurones in area 7b was decreased during *sleep*. We found (Hyvärinen and Poranen 1974, Leinonen et al. 1979) that there are neurones in area 7 that respond to the presentation of *novel visual stimuli* although they do not fire during visual fixation of familiar objects (Figs. 59 and 63). The neurone illustrated in Fig. 59 was only activated when objects were visually presented to the monkey from the left, provided the monkey had not seen these objects for some time. When raisins were presented in the correct locus and the monkey keenly followed them with its eyes no response was obtained, but each time a new object was visually presented in the correct location a clear response was recorded. Serving as such objects were various laboratory tools, a piece of paper, cotton, pencil, a flashlight, etc. Leinonen et al. (1979) found 11 neurones that responded transiently to interesting stimuli , e.g., a banana, raisin, or merely a novel object. These stimuli nearly always triggered visual fixation. After repetitive or prolonged presentations of the same object, fixation no longer resulted in a cellular response. Thus, the cellular activity was related to the interest that the object evoked in the monkey. Leinonen et al. (1979, 1980) also described neurones that fired when the experimenter's hand approached a package of raisins from which the monkey was fed (Fig. 70A). These neurones stopped responding if the animal was not given a raisin after several movements of the hand towards the package. Fixation of food did not result in activity of these neurones.

These examples indicate that motivational factors have an effect on at least some neurones in area 7 and that this effect may not always be entirely explicable on the basis of fixation or the sensory properties of the neurones studied.

Bushnell et al. (1978, 1981) emphasized the role of *attention* in the responses of area 7 neurones independent of visual fixation. They trained monkeys to fixate a target light and at the same time to attend to a peripheral visual stimulus. The responses of all area 7 neurones studied were enhanced when the monkey was attending to the stimulus in the receptive field. This enhancement was spatially specific; i.e. attention to stimuli outside the receptive field did not enhance the responses. A similar enhancement preceded a saccade towards the peripheral receptive field. However, the responses of neurones in the frontal eye fields were enhanced only when a saccade to the receptive field occured (Bushnell and Goldberg 1979, Goldberg 1980, Goldberg and Robinson 1980; Goldberg and Bushnell, 1981). These results suggest that the activity of area 7 neurones is related to sensory attention. Often the activity is also related to movement triggered by the stimulus but it can be dissociated from the movement. According to these authors the significance of the stimulus determines the response in area 7, whereas in the frontal eye fields the enhancement of cellular activity triggers an eye movement.

In recent experiments Mountcastle (1981) and Mountcastle et al. (1980, 1982) studied the effects of behavioural variables on the responses of the visual ("light-sensitive") neurones in the medial part of area 7. In one behavioural condition the monkey sat still, gazing into a uniform field without performing any task. In another condition the animal was waiting for the onset of the fixation light, and in the third condition it was performing a visual fixation task. *Interested fixation of the target light facilitated responses* to visual stimuli delivered on the receptive fields of the neurones. The response was much weaker when the same receptive field was stimulated in the absence of fixation of a target light. The facilitation effect of fixation was not explained as a general arousal effect since it was not observed when the monkey was waiting for the onset of the target light and interested in performing the task. The facilitation could be due to a selective visual attention effect or to improved optical and neural transmission caused by the fixation, or both.

E. Effects of Drugs

Eye-hand coordination is easily disturbed by drugs that impair psycho-motor performance. For instance alcohol typically disturbs eye-hand coordination and reduces the capacity to perform several tasks simultaneously (Moskowitz 1973). On the other hand, the parietal association cortex is clearly involved in eye-hand coordination, for instance in manual reaching for visually observed targets. Furthermore, as indicated in Chapter 5.1 the clinical syndrome of simultanagnosia produced by posterior parietal lesions is a defect affecting the ability to attend simultaneously to more than one task. Thus there was reason to study whether the posterior parietal association cortex is involved in the neural mediation of the effects of alcohol on sensori-motor coordination (Hyvärinen et al. 1978b). For comparison with the effects of alcohol we studied a few other drugs, too (Hyvärinen et al. 1979a, b).

These studies were made using multiple-unit recording technique with coarse microelectrodes with exposed tips of about 100 μm. In these experiments the recovery from the drug action and the return of activity to the level of the control recordings took several hours. It was not possible to hold single units at the electrode tip in non-anaesthetized, moving animals all this time; therefore multiple unit studies were performed. Since the neurones recorded close to each other appear to have similar functional properties in the posterior parietal cortex, as well as elsewhere, most neurones recorded simultaneously with a coarse microelectrode respond to similar stimuli. Thus multiple-unit recording indicates adequately the function of the neurones at the site of recording. The multineuronal impulse activity was filtered with a high-pass filter to eliminate any slow waves from the responses, rectified and integrated. The amplitudes of the integrated responses were measured as indicated in Fig. 60A. Mean and standard deviation values of ten consecutive responses were calculated. During the action of the drugs the amplitudes of the responses were expressed as percentage of the control values obtained prior to the administration of the drug. In order to obtain rapid action of the drugs they were infused intravenously through an intracardial catheter permanently implanted via the jugular vein. After washing, blood samples were taken through the same catheter for the analysis of the concentration of ethanol in central blood. The behavioural action of the drugs was monitored by observing the monkey's reaching for food rewards. Raisins or pieces of banana were moved close to the monkey in the experimenter's hand or on a stick, and the monkey's reaching accuracy and speed were scored by the experimenters. Ten points were given for rapid and accurate grasping of

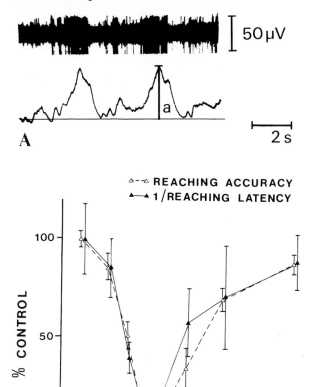

Fig. 60. A Multineuronal impulse activity (*upper trace*) and the same activity rectified and integrated (*lower trace*). Two control responses during reaching with ipsilateral arm towards a food reward are seen. Measurement of the amplitude of an integrated response from the base-line of the recording is indicated (*a*). **B** The effect of pentobarbital on the reaching accuracy score and the inverse value of the reaching latency measured electrically. At each *arrow* 4 mg/kg of pentobarbital was given through the intracardial catheter. Mean and standard deviation values of 10 trials expressed as percent of control values. (Hyvärinen et al. 1979a)

the food and 0 points for no attempt to reach. Intermediate values were given for intermediate accuracies and speeds. Three observers did the scoring, getting results which did not differ by more than ± 10%. In some experiments the latency of reaching for a food reward was separately measured using a specially constructed food well with electric timing. A close correlation was observed between the reaching scores and the inverse values of reaching latency (Fig. 60B).

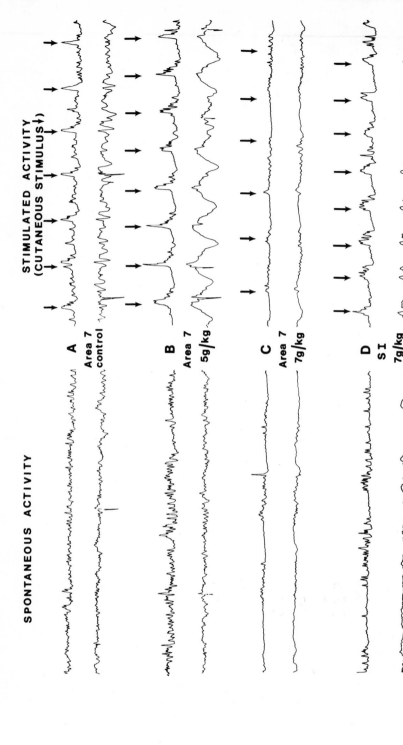

Recordings made in the lateral part of area 7 showed that the level of cellular activity in several penetrations performed at different recording sites correlated inversely with the ethanol dosage although such an effect was not observed in the primary somatosensory area. Figure 61 shows results of an experiment in which an exceptionally large dose of ethanol was given by mouth. In this recording cellular responses were elicited by cutaneous touch on the hand. The upper curve in each pair is the integrated multiple unit activity and the lower one the unfiltered focal electrocorticogram. The responses did not decrease until 7 g/kg of ethanol had been given; then the monkey suddenly became non-reactive to its environment. At the same time the responses to cutaneous stimulation also disappeared and only periodic spontaneous bursts of activity remained. We

Table 6. Effect of ethanol and pentobarbital on multineuronal impulse responses at different recording sites in area 7b. (Hyvärinen et al. 1979a)

| | Alcohol effect | | | | Pentobarbital effect | | | | | |
| | Yes | | No | | Yes | | No | | Total | |
Type of activity	N	%	N	%	N	%	N	%	N	%
Motor (reaching, grasping, manipulation)	8	67	4	33	3	37	5	63	20	32
Visual (complex visual phenomena)	5	50	5	50	11	85	2	15	23	37
Somatosensory (skin, muscle, joint)	3	30	7	70	7	78	2	22	19	31
Total	16	50	16	50	21	70	9	30	62	

The differences between the effect of alcohol and pentobarbital are statistically significant (χ^2-test) on the somatosensory responses ($P < 0.05$) and on the sensory responses (somatosensory and visual combined, $P < 0.01$). The difference between the effects on motor responses and sensory responses is significant for pentobarbital ($P < 0.05$)

◀ **Fig. 61 A-D.** Integrated multineuronal impulse activity and focal intracortical electrocorticogram at a recording site in area 7b responsive to cutaneous stimulation of distal parts of all limbs. The *left side* of the figure shows spontaneous activity and the *right side* the responses to cutaneous stimuli delivered on the contralateral hand, the site that gave the best response. Both integrated multineuronal impulse activity and focal intracortical electrocorticograms are shown. In each pair the *top record* is the integrated multineuronal impulse activity and the *lower record* the corticogram obtained through the same electrode without filtering. **A** Control recording. **B** After peroral administration of 5 g/kg of alcohol. **C** After additional 2 g/kg of alcohol. **D** The electrode was shifted to the adjacent primary somatosensory cortex (*SI*) immediately after record. **C.** Here the tissue was still strongly reactive to cutaneous stimulation of the contralateral hand although the monkey did not react to any stimulation and no responses were recorded in area 7. (Hyvärinen et al. 1978b) then quickly shifted the electrode to the adjacent primary somatosensory cortex where we still obtained strong responses to the touching of the hand (Fig. 61D).

At those recording sites in area 7 where activity was related to voluntary motor activity of the hand ethanol commonly but not always affected the cellular activity (Table 6). An example of the effect of ethanol on cellular activity and reaching accuracy is given in Fig. 62. At this recording site the cellular activity started when food was brought within the reach of the monkey and it started arm movements towards the food. The activity ceased when the monkey had the food firmly in its hand. The reaching movements became characteristically slower and clumsier with the increasing dose of ethanol and concomitantly the cellular discharge also diminished.

Under the influence of alcohol a close correlation between the amplitude of integrated multiple unit responses and the monkey's reaching accuracy was observed in 8 of the 12 locations that responded to reaching or grasping (Table 6). At the remaining four recording sites the responses were not affected by ethanol although the monkey performed very poorly in reaching for the reward. We found no features in the control responses that indicated whether a particular site was sensitive to alcohol or not.

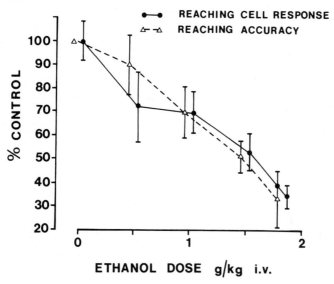

Fig. 62. The effect of the dose of ethanol on the monkey's reaching accuracy and the cellular response at a site responsive to reaching movements performed with the contralateral hand. When the monkey was reaching for a food reward the neuronal responses and the monkey's reaching accuracy were simultaneously recorded. Both fell uniformly with increasing ethanol dose. Means and standard deviations of 10 responses expressed as a percentage of the mean values of control record. (Hyvärinen et al. 1978b)

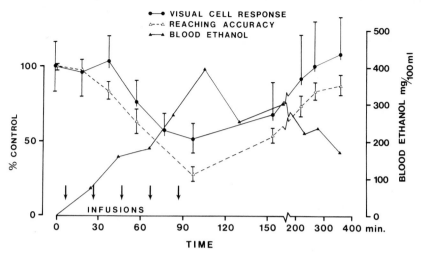

Fig. 63. Effects of ethanol on cellular response, the monkey's reaching accuracy and blood ethanol level in an experiment performed at a site in area 7 responsive to the introduction of new objects into the monkey's field of vision. The arrows indicate intracardiac infusions of 0.5 g/kg of alcohol each giving a total of 2.5 g/kg. The integrated multineuronal impulse responses and the monkey's reaching accuracy (mean ± SD of 10 responses) are expressed as a percentage of mean control values. Ten different laboratory tools and other objects not previously seen by the monkey and conveniently available were successively presented at each phase of the experiment. Repeated presentations of the same stimulus object, even food that triggered visual fixation, did not activate these cells. (Hyvärinen et al. 1978b)

Visual sensory responses in area 7 were also affected by ethanol. Figure 63 shows an example of the effect of alcohol on a complex visual response. At this recording site cellular activity was triggered by new objects that were brought into the monkey's visual field. Such objects as various laboratory tools were used, ten of which were consecutively presented to the monkey to obtain the mean and standard deviation values shown in the figure. Fixation of food did not activate this site nor did repeated presentations of the same object. As can be seen in the figure, the responses diminished with increasing blood alcohol concentration and recovered during the elimination of alcohol. The amplitudes of the responses also correlated with the reaching accuracy of the monkey. A difference between the effects of rising and falling alcohol concentrations was observed. In Fig. 63 a blood concentration of 200 mg/100 ml during the rising phase of intoxication reduced the cellular responses and reaching accuracy nearly by half, whereas during the recovery phase almost normal cellular and behavioural responses were observed at this level of blood ethanol, indicating an acute tolerance effect.

At five of the ten visual recording sites alcohol significantly reduced the responses, whereas at the other five sites with this type of response no effect of ethanol on the response was observed (Table 6). At the latter recording sites the cellular responses remained unchanged even when the monkey was heavily intoxicated and did not seem to attend to the stimuli. Neither could we predict on the basis of the control recording whether the responses of the visual recording sites would be influenced by alcohol or not.

The effect of ethanol was observed at one half of the recording sites in area 7, whereas with the dose (up to 2 or 3 g/kg) commonly used the other half of the recording sites remained uninfluenced (Table 6). The effects were most common at the recording sites where the activity was related to motor behaviour of the monkey, less common at recording sites characterized by responses to complex visual stimuli and least common at the recording sites with somatosensory responses. However, with a larger dose of ethanol more of the somatosensory sites were affected, whereas no influence was observed in the primary somatosensory cortex. Thus, area 7 is far more sensitive to alcohol than SI. It is well known that general anaesthesia does not block the responses in the primary somatosensory cortex, whereas it does so in the association areas (Albe-Fessard and Fessard 1963, Thompson et al. 1963).

It has been suggested that the effect of alcohol depends on the complexity of synaptic contacts to the site of recording (Himwich and Callison 1972, Kalant 1975). On the basis of EEG and evoked potential studies the cortical association zones and the reticular formation of the brain stem, a polysynaptic structure, were reported to be quite sensitive (DiPerri et al. 1968, Perrin et al. 1974). Our findings in the parietal association area 7 support this interpretation since no effects were observed in the primary somatosensory cortex, and within area 7 the effect correlated with the complexity of the response properties. However, these results indicate that cortical loci that appear functionally similar may have different synaptic contacts from the pharmacological point of view.

It has commonly been assumed that the effects produced by alcohol are non-specific and similar to the effects produced by other centrally acting depressant drugs such as barbiturates (Kalant 1975). Therefore we compared the effects produced by ethanol in area 7 to those produced by pentobarbital (Hyvärinen et al. 1979a). However, for such a comparison it is important to administer both drugs in doses with comparable behavioural effects. If anaesthetic doses of pentobarbital are used the results are certainly different from those obtained with ethanol. Therefore, we used relatively small doses of pentobarbital that produced behavioural incoordination comparable with that produced by ethanol. The doses needed to produce similar effects with both drugs were adjusted by estimating the monkey's reaching accuracy for food reward using the scale mentioned above.

The results of the pentobarbital and alcohol experiments are presented for comparison in Table 6. Pentobarbital affected motor responses less frequently than alcohol: in the former group 3 of 8 sites were affected, whereas 8 of 12 sites were sensitive to alcohol. This difference was not statistically significant, however. Sensory responses, on the other hand, were more often affected by pentobarbital than by alcohol, and this difference was statistically significant.

These results indicate that the effects of ethanol and small doses of pentobarbital are similar at some recording sites in area 7 but that there are statistically significant differences in the distribution of the effects of the two drugs. *Whereas ethanol more commonly affects cells whose activity is related to motor behaviour, the barbiturate, even in small doses, affects the sensory responses significantly more often.* Such a difference could be expected on the basis of the difference in the behavioural action of the two drugs. Although small doses of pentobarbital produce a general appearance of sedation and a tendency to fall asleep, the motor performance is not much disturbed when awake. Alcohol, on the other hand, produces often a euphoric state during which motor behaviour increases but becomes incoordinated. The difference in the observed distribution of effects is consonant with the behavioural action of the two drugs. The general sedation produced by pentobarbital correlates with its depressing action on the sensory responses in area 7, whereas the motor incoordination produced by alcohol correlates with the effect that this drug has on the cellular discharges related to motor performance.

From the point of view of the association cortex the so-called "dissociative anaesthetics" are of interest as they are thought to block the sensory inflow to the association systems of the brain. Phencyclidine was originally developed as a veterinary tranquillizer (Domino 1964) but it is nowadays also abused in the streets under the names PCP or "angel dust" (Liden et al. 1975, Lemberger and Rubin 1976). Its congener ketamine is used in human surgery for anaesthesia in minor operations. During induction of ketamine anaesthesia the recipient is said to feel dissociated from his environment including his own extremities (Goodman and Gilman 1970). This symptom recalls the somatic symptoms asomatognosis, autotopognosia and asymboly for pain that are produced by posterior parietal lesions (see Chap. 5.1). In a study of the effects of phencyclidine on area 7 we found that its effects were much stronger and quite different from those of other drugs (Hyvärinen et al. 1979b).

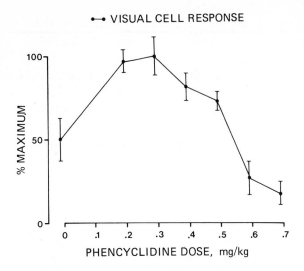

•—• VISUAL CELL RESPONSE

Fig. 64. Effect of phencyclidine on the amplitudes of integrated responses to visual stimuli at a locus in area 7 responsive to the approach of objects toward the face of the monkey. Mean and standard deviation values of 10 responses. (Hyvärinen et al. 1979b)

With small doses (less than 0.5 mg/kg) phencyclidine produced symptoms typical of this drug including deterioration of motor control, wandering gaze, and mastication. At this phase the amplitudes of the integrated responses to sensory stimuli started to increase (Fig. 64). With larger doses the animal stopped reacting to external stimuli including pain, and the cellular responses declined. At a dose of 1 mg/kg the responses were considerably weaker than in the control recordings. At still larger doses there were periods of complete lack of cellular activity.

The two phases of the effect of phencyclidine correlate with the concurrent behavioural effects. In humans, small doses are known to produce euphoria, hallucinations and thought disorders (Domino 1964, Lemberger and Rubin 1976), whereas larger doses produce anaesthesia. Actions of phencyclidine are likely to occur at other parts of the brain, too. However, the above results indicate that phencyclidine, at sufficient dosage, blocks sensory inflow to the parietal association cortex, and is thus confirmed to act as a "dissociative" anaesthetic.

These studies of the effects of drugs on the posterior parietal cortex indicate that this associative cortex is quite sensitive to pharmaca. The actions of various drugs on this cortical network are not, however, straightforward, since there were many recording sites where no effects were observed. Whether an effect is present or not may depend on the different types of connections, transmitters, and metabolic factors operative in the various cellular systems. However, when the normal cellular action of the posterior parietal cortex is studied the experiments must be performed using chronic recording technique which does not necessitate the use of anaesthesia or tranquillizers that could contaminate the results with drug actions.

IX. Vestibular and Auditory Responses in the Parietal Lobe

A. Vestibular Responses

A vestibular projection to the parietal lobe was demonstrated in a separate cytoarchitectural area called 2v which lies at the lateral end of the intraparietal sulcus at the junction of areas 2, 5 and 7b. Neurones in this region also respond to proprioceptive somatosensory stimuli and to optokinetic visual stimuli (Schwarz and Fredrickson 1971b, Büttner and Buettner 1978). About half of the vestibular neurones in this cortical field were activated during chair rotation to the contralateral side, whereas rotation in the opposite direction inhibited them. Optokinetic stimulation excited most of these neurones; as expected, the effective direction of rotation of the visual surround was opposite to the effective direction of head rotation. Some of these neurons were also activated by passive limb movements or muscle tapping, and occasionally activity increased during intended head movements (Büttner and Buettner 1978).

It is thus clear that area 2v is a polysensory region that receives vestibular, proprioceptive and visual inputs. Together these mechanisms may contribute to the sensory analysis and motor control of the body position in relation to the environment during movements.

This area lies at the junction of the primary somatosensory and the associative posterior parietal cortex. Cytoarchitecturally it has been considered a subregion of area 2, which implies that it has primarily a somaesthetic function. Although connections from area 2v to the posterior parietal lobe have not been separately analyzed, 2v is likely to project to areas 5 and 7 as other parts of area 2 do (Selzer and Pandya 1980) (see Chap. IV). Vestibular input from the thalamus (Deecke et al. 1977, Hawrylyshyn et al. 1978) to the posterior parietal cortex is also possible. Such projections could give a vestibular contribution to the posterior parietal sensory and motor control mechanisms.

It is now well established that neurones in the vestibular pathways also respond to optokinetic stimuli. Responses to optokinetic stimuli were elicited in the vestibular nuclei (Henn et al. 1974, Keller and Kamath 1975, Waespe and Henn 1977), in the thalamus (Büttner and Henn 1976),

and in the cortex (Büttner and Buettner 1978). In humans rotation of the
visual surround produces the sensation of circular vection which is indis-
tinguishable from true body rotation (Young et al. 1973). Moreover, the
vestibular system is important for the coordination of head and eye movements during smooth visual pursuit (Lanman et al. 1978). Since many of
these mechanisms affect the function of the posterior parietal lobe, it is
likely that the vestibular system influences the activity of the posterior
parietal neurones. However, the testing of vestibular input with natural
stimuli requires another type of experimental set-up with the animal
mounted on a turntable. Therefore, such studies have only recently been
performed.

Kawano et al. (1980) showed that visual tracking neurones recorded in
the posterior part of area 7 are activated by vestibular stimulation. In a
small sample of neurones active during horizontal visual tracking several
were found to respond to horizontal rotation of the animal in complete
darkness. Most of them responded during head rotation to the ipsilateral
side which was also the direction of visual tracking that activated these
neurones. Moreover, when the monkey was drowsy and the vestibulo-
ocular reflex extinguished, a similar response to rotation was observed in
the dark. Therefore, the vestibular responses were genuine and not mediated by the eye movements caused by vestibular stimulation.

Since the preferred directions of visual tracking and head rotation
were the same for these neurones, area 7 does not appear directly involved
in the vestibulo-ocular reflex. Although this finding is so far based on a
small number of observations, it rather suggests that the posterior parietal
mechanisms could serve the process of reorganization of the relationship
between eye and head movements when visual tracking or search is performed during head movements. They could participate in a postulated
mechanism of an internal smooth pursuit command driving both eye and
head movements (Lanman et al. 1978).

B. Auditory Responses in Area Tpt

In the parietal association cortex of chloralose anaesthetized cats
responses to auditory stimuli can be recorded using evoked potential
technique (Thompson et al. 1963). We have often tested, in connection
with other studies, for auditory responses in area 7 of awake monkeys
using both natural and synthetic auditory stimuli but so far we have not
observed auditory responses in area 7 proper. However, ablation of the
temporo-parietal association cortex of the monkey produces inattention
towards contralateral auditory stimuli (Heilman et al. 1971), and human

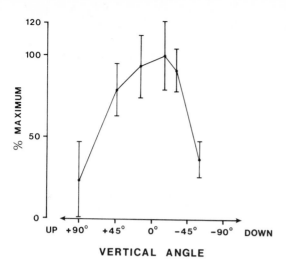

Fig. 65. Amplitudes (means and standard deviations of 10) of integrated responses in a group of neurones in area Tpt to auditory stimuli presented at different angles in the horizontal and vertical interaural planes. The maximum integrated response is indicated as 100%. The stimuli were produced by hitting together two small metal bars at a distance of about 40 cm from the monkey. (Hyvärinen 1981a)

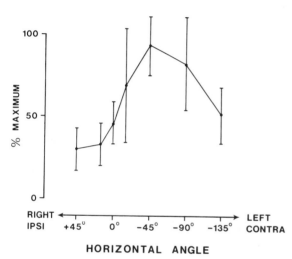

patients with lesions in this region neglect contralateral auditory stimuli (Heilman and Valenstein 1972a). Central in this cortical region is area Tpt (Fig. 3) as defined for the monkey by Pandya and Sanides (1973) and for man by Galaburda and Sanides (1980). Pandya and Sanides consider area Tpt cytoarchitecturally akin to area 7. In humans area Tpt occupies a comparable region and extends to the most lateral and posterior part of the parietal lobe (Galaburda and Sanides (1980). Lesions in this part of the parietal lobe have produced indifference to loud and unpleasant noises (Schilder and Stengel 1931).

This background knowledge suggested that area Tpt might respond to auditory stimuli and resemble otherwise area 7. Therefore, we investigated the responses of neurones in area Tpt using acoustic, visual and somatic stimuli (Leinonen et al. 1980). Initial mapping experiments performed with multiple-unit recording indicated that acoustic stimuli were effective in this cortical region. Furthermore, they showed that neurones in this region were *sensitive to the location of the sound source.* An example is given in Fig. 65, which shows the amplitude of integrated multiple unit activity of a group of neurones in response to auditory stimuli. These cells responded to pure tones between 5000 and 16000 Hz. The best responses were obtained with click sounds produced by hitting together two small metal bars on the contralateral (left) side of the monkey. The upper curve shows the influence of the vertical angle of the stimulus on the response and the lower one the influence of the horizontal angle. The maximal response was obtained approximately at the level of the monkey's ears and about 40° to 60° to the left from midline. The distance of the stimulus from the monkey's head was approximately 40 cm (Hyvärinen 1981a).

The single-unit recordings in area Tpt (Leinonen et al. 1980) were made using acoustic stimuli in free field in addition to visual and somatic stimuli of the same type as in previous studies. Pure tones with frequencies from 0.2 to 20 kHz, clicks, band-limited white noise, and different natural sounds produced by the experimenters and analyzed spectrographically were used. The effects of the movement of a sound were studied by moving the sound source by hand or by a sound movement simulator.

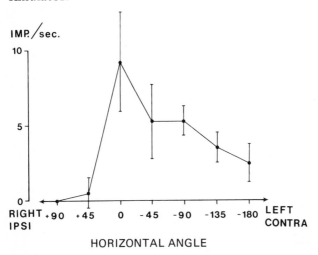

Fig. 66. The discharge rate of an area Tpt neurone as a function of the horizontal angle of incidence of the sound. A rattling sound was presented at various locations in the horizontal interaural plane at a distance of 50 cm from the monkey. Means and standard deviations were calculated from 10 presentations in each location. The best response was to sounds produced straight in front of the animal. (Leinonen et al. 1980)

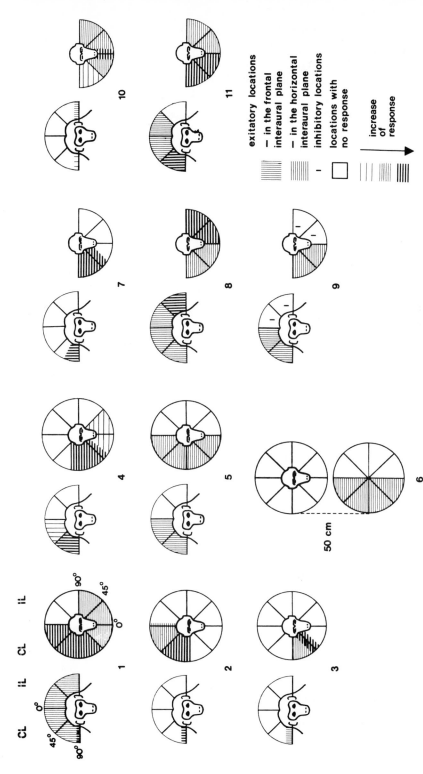

Fig. 67. The effective locations of the sound source in the frontal and horizontal interaural planes for eleven neurones in area Tpt. The stimuli (various natural sounds) were delivered at a distance of 50 cm from the head. Cell No. 6 responded only when the source was about 50 cm below the horizontal interaural plane. (Leinonen et al. 1980)

About 70% of the neurones recorded in area Tpt responded to acoustic stimulation. Most cells were activated by natural sounds. Rubbing of fingers, hands or a cloth were effective stimuli. These sounds are only slightly above the monkey's threshold of hearing, their spectrum is noisy with most audible energy between 2 and 20 kHz and contains rapid intensity and frequency modulations. Many neurones also responded to human speech sounds, whereas monkey vocalizations were ineffective. About half of the cells examined responded to pure tone stimuli, most of them over a wide frequency range (3 to 5 octaves). Most cells tested responded to clicks either with excitation or inhibition, whereas most responses to white noise were inhibitory. The response latencies were typically between 15 and 60 ms but they were quite variable.

The location of the sound source in the auditory space influenced about 70% of the neurones studied. The strongest response was usually elicited by sounds coming from a particular direction, usually from either side. An example is given in Fig. 66. Most neurones responded to stimuli in both halves of the auditory space, but about half on them were activated more by sounds coming from the side contralateral to recording. A few cells were best activated by sounds presented on the ipsilateral side. Figure 67 shows examples of the effective locations of the stimuli for 11 neurones in area Tpt.

Close to 40% of the neurones in area Tpt were activated by somatic mechanisms. Most neurones responding to cutaneous stimulation had a receptive field around or on the ear lobe; usually they responded to stretching of the ear lobe or to blowing into the hair around the ear. During such studies the monkey's ears were covered or the sounds produced by the somatic stimuli were masked with a continuous loud noise from a loudspeaker. The locations of the receptive fields for neurones responding only to cutaneous stimulation are shown in Fig. 68A. Three cutaneously drivable neurones also responded to rotation of the head; they may have received proprioceptive input from the neck or vestibular input. A few neurones were activated only by such stimuli.

Fig. 68 A-C. Properties of neurones in area Tpt activated by somatic mechanisms. ▶
A Locations of cutaneous receptive fields of the neurones responding only to somaesthetic stimuli. **B** Locations of somaesthetic receptive fields and effective locations of the sound source in the horizontal interaural plane at a distance of 50 cm from the monkey for neurones that responded to both somaesthetic and auditory stimuli. Four neurones at the right responded to pure tones and discharged during hand or head movements; no somaesthetic receptive fields were found for these neurones. **C** Cutaneous receptive field and the effective direction of visual stimulus movement in one cell. (Leinonen et al. 1980)

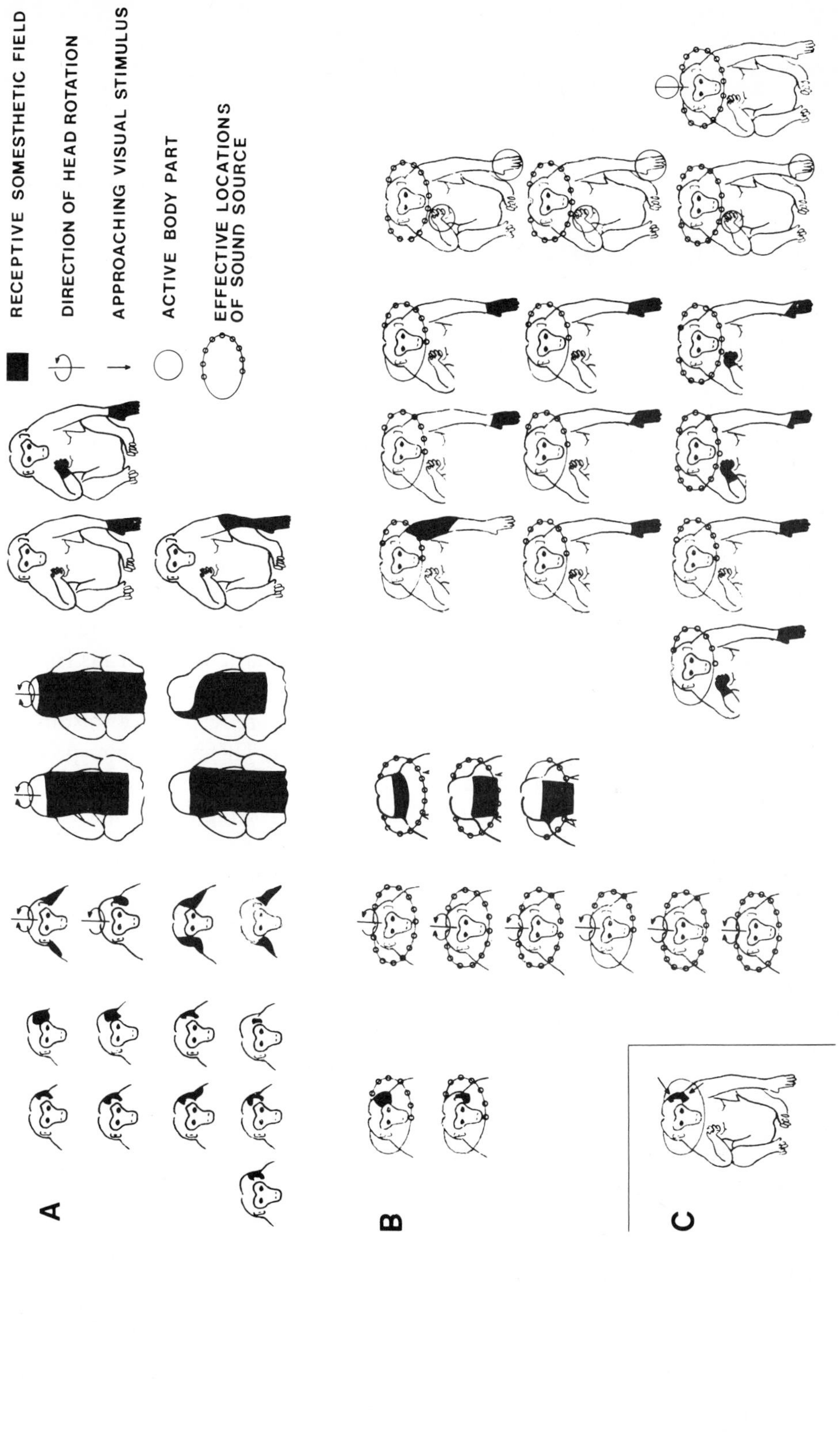
RECEPTIVE SOMESTHETIC FIELD
DIRECTION OF HEAD ROTATION
APPROACHING VISUAL STIMULUS
ACTIVE BODY PART
EFFECTIVE LOCATIONS OF SOUND SOURCE
A
B
C

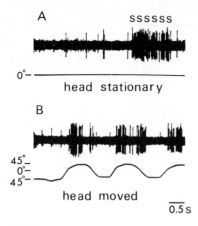

Fig. 69 A, B. Responses of a neurone in area Tpt to auditory stimulation and head rotation. **A** Auditory response to a hissing sound (ssss) produced behind the monkey on the contralateral side. **B** Responses to passive head rotation toward the contralateral side. The *curve* below the cellular discharge indicates head rotation upwards deflection indicating movement of the face towards the contralateral side. (Leinonen et al. 1980)

Ten percent of the neurones recorded in area Tpt responded both to somaesthetic and auditory stimuli (Fig. 68B). The effective somaesthetic stimuli for these neurones were similar to those mentioned above, i.e., tapping or touching the skin around the ear, blowing into the hair on the neck or shoulder, passive rotation of the head, and in addition tapping of the fingers. All the neurones that had only contralateral somaesthetic receptive fields were activated by sounds only from the contralateral side, whereas those neurones that were activated by somatic mechanisms from both sides were also activated by auditory stimuli from both sides. Passive head rotation also activated some neurones that were responsive to sounds. An example of such a neurone is shown in Fig. 69. This neurone responded to white noise in the contralateral auditory space and passive head rotation. The latter response was not influenced by covering the monkey's ears.

A few neurones were also observed to respond to both auditory and visual stimuli. Examples of the responses of three such neurones are shown in Fig. 70.

The above results show that several of the functional properties of the neurones in area Tpt resemble those of neurones in area 7 proper. Similar visual and somaesthetic stimuli activated neurones in area Tpt as in area 7. Differences between these areas concerned three functional properties. First, the location of the somaesthetic receptive fields in area Tpt was most commonly around the ears or in the head and neck region, whereas in the other parts of area 7 they were on the hands, arms, and face or less frequently on the trunk or legs. Secondly, the motor mechanisms present in area Tpt were related to head movements, whereas in other parts of area 7 they were related mainly to eye, arm, and hand movements. Thirdly, the neurones in area Tpt responded to auditory stimuli; such neurones have not been found in other parts of area 7.

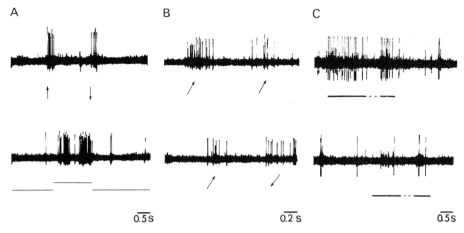

Fig. 70 A-C. Responses of three different neurones in area Tpt to visual and auditory stimuli. **A** above, the cell responds when the monkey looks at the experimenter's hand approaching a package of raisins (↑) and when the hand is withdrawn (↓); below, the cell responds when a rattling sound continuously present in the contralateral half of the auditory space moves (positive deflection of the timing pulse). **B** above, the cell responds when the monkey is looking at a small rotating cuboid box, the cell always discharging during certain phases of the rotation (↗); below, the cell responds when a 10 kHz tone is moved behind the monkey toward it (↗) or away from it (↘), with the sound continuously present. **C** above, the cell discharges when a sheet of paper is brought into the contralateral periphery of the visual field from behind the monkey (——), becomes inactive when the stimulus is stationary (– – –) and discharges again when it is taken away (——); below, the response to the same visual stimulus is inhibited by a 10 kHz tone which is continuously on. (Leinonen et al. 1980)

Area Tpt does not receive direct auditory input from the medial geniculate nucleus. However, auditory afferents to area Tpt arrive from the parakoniocortex (paAlt) surrounding the auditory region in the supratemporal plane (Pandya and Sanides 1973). This region is also known to project to the auditory pathway: medial geniculate nucleus, inferior colliculus, and nucleus paralemniscalis (Kuypers and Lawrence 1967).

The effective auditory stimuli for neurones in area Tpt were various natural sounds such as rubbing of hands, rustle of clothes, clicks produced by hitting objects against each other, human speech, etc. The frequency of pure tones or the sound pressure levels did not appear very important; proper stimulus features and sufficient signal-to-noise ratio (acoustic contrast) were the determinant factors. Moreover, the sensitivity of these neurones to such stimuli was phenomenal; often responses were elicited by quite distant stimuli well below human auditory threshold. In the auditory associative zone CM neurones resemble those we found in area Tpt responding to pure tones over a range of several octaves at threshold sound pressure levels and preferring weak sounds (Brugge and Merzenich 1973,

Merzenich and Brugge 1973, Imig et al. 1977). In their natural habitat, the tropical forest, monkeys orient themselves in the midst of faint rustles of leaves and communicative sounds of other animals. Under such conditions the capacity to orientate rapidly on the basis of faint auditory signals is often crucial for survival. Thus, from an evolutionary point of view the sensitivity of these neurones to faint natural sounds is well understandable.

The character of the sounds stimulating these neurones also resembles the auditory illusions produced by electrical stimulation of the temporal lobe in man (Penfield and Rasmussen 1950). In the associative temporal cortex electric stimulation does not cause perception of pure tones or other acoustically elementary sounds but illusions that people are moving, furniture colliding, doors closing, etc. Thus in associative auditory areas the absolute physical properties of the stimulus sounds may be of secondary importance, whereas their communicative significance may play a more important role than in the auditory pathways and the primary auditory cortex.

An important feature of the acoustic sensitivity of area Tpt neurones is their sensitivity to the angle of incidence of the sound. This sensitivity may be rather independent of the spectral properties of the stimuli. Brugge and Merzenich (1973) showed that in the primary auditory cortex (AI) of the monkey both the spectral properties and the interaural intensity and phase differences are important for the activation of the neurones. Moreover, a tonotopic organization appears well established in AI (Merzenich and Brugge 1973). However, in area Tpt the mechanisms related to sound source localization appeared more important than those related to spectral properties. The crude nature of our acoustic stimuli delivered in free field in a noisy laboratory precludes us from negating the existence of highly specific pitch-related mechanisms in area Tpt, but if they exist they are probably related to the spectral properties of the sounds that are useful in the localization of their souce (Whitfield 1971).

It is not inconceivable that area Tpt, or some other parts of the associative auditory cortical regions, is organized to represent various parts of the auditory space. Such an organization would mean a change from the tonotopic organization prevailing in the auditory pathway and primary cortical areas to a "spatiotopic" organization. An interesting conversion from tonal to spatial representation in the auditory system has been described in the receptive fields of neurones in the midbrain of the owl (Knudsen and Konishi 1978).

The somatomotor mechanisms in area Tpt were prominently related to movements of the head. Electric stimulation studies have also yielded results that are consonant with this finding; stimulation of the cortex around the caudal end of the Sylvian fissure produced movements of the ears (von Bechterew 1911, Vogt and Vogt 1919) and of the ears and head (Lilly 1958). The association of auditory activation with head movements does not appear accidental. Head turning towards interesting sounds aids their localization and is a central feature in the classical orienting reflex, whereas movements of other body parts have no significance for auditory localization. The auditory inattention or neglect produced by lesions in the temporo-parietal cortex (Heilman et al. 1971, Heilman and Valenstein 1972a) may thus reflect damage to mechanisms controlling head movements and direction of attention towards the contralateral auditory space.

X. Regional Distribution of Functions in Area 7

Many of the results referred to above suggest that within area 7 there are *functionally different regions*. Electric stimulation studies indicated a difference between areas 7a and 7b; the Vogts (1919) showed that stimulation of area 7a produced eye movements whereas stimulation of area 7b produced hand movements (Fig. 38). The microelectrode recording studies of Leinonen et al. in my laboratory (Leinonen et al. 1979, Leinonen and Nyman 1979, Leinonen 1980) showed that the lateral part of area 7 (7b) contains more somatic mechanisms than the studies of Mountcastle et al. (Mountcastle et al. 1975, Lynch et al. 1977, Motter and Mountcastle 1981) which were performed mainly in area 7a and were dominated by visual and oculomotor mechanisms. In a preliminary mapping study we showed that visual-oculomotor mechanisms were concentrated in the medial part of area 7, whereas somatic mechanisms were more common laterally and posteriorly (Hyvärinen and Shelepin 1979). Furthermore, most laterally in area 7b in the parietal operculum and in the upper lip of the Sylvian fissure Robinson and Burton (1980b,c) described purely somaesthetic responses.

It thus seemed obvious that there were differences in the functional properties of neurones in different parts of area 7. In order to find out possible grouping of the functionally different response types in area 7, I made a systematic microelectrode mapping study along the whole convexity of the medial parietal gyrus covering both areas 7a (PG) and 7b (PF). However, the deeper parts of area 7 buried in the sulci were not included in this study because their investigation requires a different anatomical approach. Because of the blocking effect that anaesthetics have on the neuronal activity in this region this mapping study, unlike mapping studies in general, had to be made in non-anaesthetized, behaving animals (Hyvärinen 1981c).

A. Mapping Methods

Maps were constructed of five hemispheres of three monkeys. The recordings were made with glass-covered tungsten microelectrodes with uninsulated tips of 50 to 100 μm and AC impedances of 20 to 100 kohms. Usually such electrodes recorded the activity of several neurones simultaneously (see Fig. 60A). Each electrode penetration was marked on a chart consisting of the 1 x 1 mm surface coordinate grid used in connection with the Evarts' (1966) microdrive. Only the first 2.5 mm of the penetration from the top of cortex were included in the study.

At each recording site the activity was identified with the following tests: a thorough examination of all body surfaces for detection of receptive fields on the skin, manipulation of joints and muscles, visual presentation of sheets of cardboard with patterns of various shapes and colours, presentation of flashing lights, moving food or other objects towards or away from the animal, showing new objects not previously seen by the animal, and masking and unmasking the animal's view. Furthermore, the monkey was enticed to perform active movements with its hands, arms, legs, mouth and eyes towards food and other objects. At each penetration site the functional properties of the neurones were documented in the protocol at the first depth where these properties could be well defined.

For the construction of the maps the functions at each locus studied were somewhat arbitrarily classified to seven groups as follows:

1. *Visual;* discharges elicited by various types of visual stimuli, independent of eye movements but related to stimulus movement.

2. *Oculomotor;* a reliable relation between the discharge and eye movements of stimuli.

3. *Cutaneous;* discharge triggered by mechanical stimulation of the skin.

4. *Joint;* discharge produced by passive movement of one or more joints.

5. *Muscle;* discharge produced by passive palpation or tapping of muscles. Discharge was often also observed during active movement.

6. *Somatomotor;* discharge produced by active movement of a body part independent of external stimuli.

7. *Active touch;* discharge that was slight and inconsistent when the hand was cutaneously stimulated but strong and reliable during active manipulation of objects.

The locations of the various responses were transferred to maps of the right area 7 as explained in the original paper (Hyvärinen 1981c). For quantitative evaluation of the differences in the distribution of various functions on the surface of area 7 two arbitrary coordinate lines were drawn perpendicular to each other (Fig. 71). Differences between the mean x; or y-coordinate values were evaluated using Student's t-test.

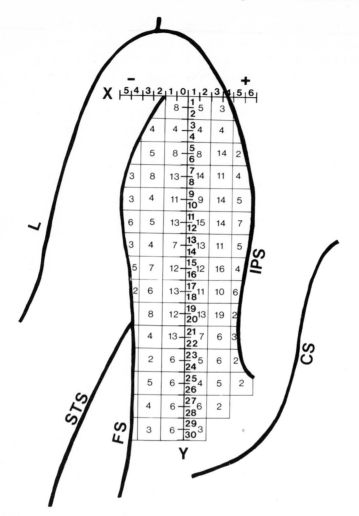

Fig. 71. The x- and y-coordinates for measurement of the recording sites in area 7. The x-axis was placed at the medial border of area 7 with positive coordinate values anteriorly and negative ones posteriorly. The y-axis was placed in medio-lateral direction along the medial parietal gyrus. The coordinate numbers are in millimeters. The *numbers* inside each 2 x 2 mm square give the number of recording points studied in that part of the area. (Hyvärinen 1981c)

As Fig. 71 indicates the sample of recording points in area 7 was not homogenous in all exposed parts of area 7. In the central part of the area there were more observations than closer to the edges. Part of the variability was caused by uneven sampling from different parts of area 7 in each hemisphere. Much of the variation was also caused by the superimposition of the sulcal patterns of different hemispheres in the construction of the maps. This procedure led to the concentration of points in the central region of the map leaving the edges less densely represented since the variability in individual hemispheres accumulated at the edges of the region studied. In principle it would be more advantageous to make complete maps of each individual hemisphere, but with the transdural recording technique that I used in conscious, behaving animals this was not technically possible. At this stage we must therefore accept the unavoidable variation caused by the superimposition of the results from several hemispheres. If it is possible in the future to carry out complete mapping of an individual hemisphere, the regional separation between functions may be even more clear-cut than shown by the distributions presented below which were widely overlapping.

B. Distribution of Responses

1. Visual Responses

Cellular responses related to only visual stimulation or eye movements were concentrated in the medial part of area 7, i.e. area 7a or PG (Fig. 71A). Altogether 122 such points were encountered (Table 7), and in most of them the cellular discharge was related to movement of the stimuli rather than to movement of the eyes.

Table 7 indicates the means and standard deviations of the x- and y-coordinates of the various recording points. The mean of the visual group was closest to the medial border of area 7. The map, as well as the statistical analysis, showed that this group of responses was a clear entity differing highly significantly along the y-axis from all other groups ($p < 0.001$, t-test).

In connection with this study and including data points from other studies the occurrence of *ocular dominance groups* in area 7 was determined. For this purpose the responses to visual stimuli were separately evaluated through each eye, keeping the other eye covered during the test. It has generally been assumed that neurones in area 7 respond equally to stimulation of each eye; this was found to be true at least on the multiple unit recording level (Fig. 73). Most recording sites (93%) were influenced equally through both eyes, 6% with a slight dominance of the contralateral eye and 1% (three sites) with exclusive activation from the contralateral eye. All the cell groups activated only through the contralateral eye had their visual receptive fields in the monocular crescent in the farthest contralateral part of the visual field. No cell groups with ipsilateral dominance were observed in area 7.

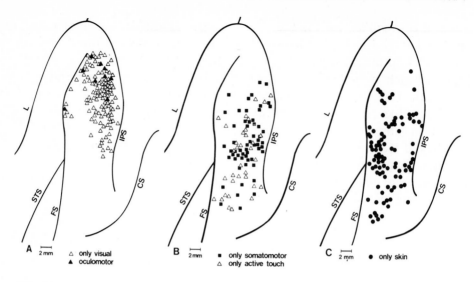

Fig. 72 A-C. A The distribution of the recording points which were found to be only visual or oculomotor in nature. **B** The distribution of the points where the activity was related only to somatomotor actions or to active touch. **C** The recording points where only cutaneous stimulation was effective. (Hyvärinen 1981c)

Table 7. Distribution of points with various response types along the x- and y-axes of Fig. 71. (Hyvärinen 1981c)

Function	n	Mean y, mm	S.D.	Mean x, mm	S.D.
1. Pure responses	359	13.4	6.9	0.8	2.6
Visual only	122	7.9	4.4	1.5	2.4
Cutaneous only	95	17.7	6.3	0.1	2.7
Kinaesthetic only	51	12.2	5.1	-0.9	2.5
Active touch only	34	18.8	6.2	0.8	2.4
Somatomotor only	57	15.9	5.4	1.7	2.3
2. Combined responses	176	16.4	6.4	1.6	2.2
Visual and cutaneous	35	20.7	4.2	1.5	1.9
Visual and somatomotor	68	13.2	4.5	1.5	2.1
Visual and active touch	18	13.9	6.5	1.2	2.8
Cutaneous and somatomotor	14	22.2	3.8	2.4	2.3
Visual and cutaneous and somatomotor	19	22.7	4.8	2.1	2.3
Visual and somatomotor and active touch	22	12.3	5.7	1.5	2.1
3. Pure and combined responses together	735	15.9	6.9	1.2	2.4
Visual all	284	13.3	7.0	1.4	2.2
Cutaneous all	178	19.3	5.9	0.6	2.6
Active touch all	93	16.9	6.7	0.8	2.4
Somatomotor all	180	15.9	6.0	1.6	2.2

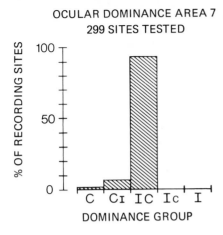

OCULAR DOMINANCE AREA 7
299 SITES TESTED

% OF RECORDING SITES

DOMINANCE GROUP

C   Cɪ IC Ic   I

I = IPSILATERAL , C = CONTRALATERAL

Fig. 73. Ocular dominance distribution of visually evoked multiple unit responses in area 7 determined at 299 recording points. *C* contralateral eye; *I* ipsilateral eye; *large letter* means strong influence, *small letter* means weak influence

Regarding ocular dominance area 7 thus differs from the visual cortex where all different ocular dominance columns are found (Hubel and Wiesel 1968, 1977). Apparently the visual functions of area 7 are such that *both eyes affect evenly most neurones* there; the only exception was a few neurones serving the contralateral monocular crescent.

2. Somatic Responses

Figure 72C shows the distribution of the recording sites that responded only to mechanical stimulation of the skin. On the average they were located 10 mm more laterally than the visual points (Table 7); this difference was highly significant (p < 0.001, t-test).

The construction of the maps revealed that responses elicited by passive rotation of joints and palpation of muscles were concentrated in the posterior part of area 7a (Fig. 74A). In the antero-posterior (x-axis) direction the distribution of the kinaesthetic points differed highly significantly from that of the visual only, visual all, and somatomotor points (Table 8).

The finding of a kinaesthetic zone in the posterior part of area 7a was unexpected, because no such activity has been previously described here. This region was also devoid of visual representation; thus here the kinaesthetic input was "modality-pure", i.e. only activity arising from passive stimulation of joints and muscles was effective. These functions were separately represented here as if this region constituted a somaesthetic pro-

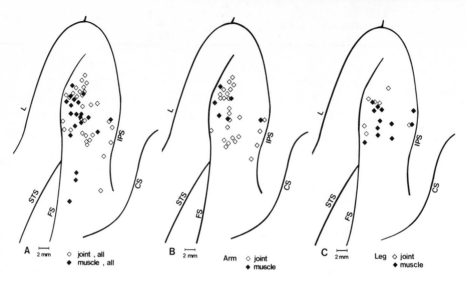

Fig. 74 A-C. Distribution of the kinaesthetic responses evoked by passive rotation of joints and palpation of muscles. **A** All points with joint or muscle responses. **B** The joint and muscle responses evoked from the arm. **C** The joint and muscle responses evoked from the lower limb. (Hyvärinen 1981c)

Table 8. x-axis distribution of kinaesthetic points (Fig. 74A) and the significance of their difference from visual only (Fig. 72A), visual all (Fig. 76A) and somatomotor all (Fig. 76B) distributions. (Hyvärinen 1981c)

	n	Mean x, mm	S.D.
Kinaesthetic	51	-0.9	2.5
Visual only	122	1.5[a]	2.4
Visual all	284	1.4[a]	2.2
Somatomotor all	180	1.6[a]	2.2

[a]$p < 0.001$, Student's t-test

jection area without visual representation, whereas in the region of somatic movements visual and somatic activity occurred together.

The distribution of active touch responses, evoked when the monkey actively manipulated objects (Fig. 72B) differed highly significantly from that of the kinaesthetic representation ($p < 0.001$), but it largely coincided with the cutaneous representation (Table 7). Responses considered purely somatomotor were distributed largely similarly to active touch responses, their distributions differing almost significantly ($p < 0.05$) in the y-axis dimension.

3. Combined Responses from Several Modalities

At 176 points responses were evoked by more than one sensory modality or type of motor action. These combinations were divided into six classes (Table 7, group 2). Activation elicited by both visual and cutaneous stimulation at the same recording site occurred in the lateral part of area 7 (7b), but was absent at the top of the gyrus in area 7a (Fig. 75A). Combination of visual and somatomotor responses (Fig. 75B) occurred in the middle of area 7 extending to both area 7a and 7b. The same medial region was also covered with points activated by visual stimuli and active touch (Fig. 75B).

Combinations of three different response types were also observed. Statistical testing (Table 9) indicated that all three combinations that included the skin (Fig. 75A) were located significantly more laterally than the combination of visual responses with somatic movement and active touch (Fig. 75B).

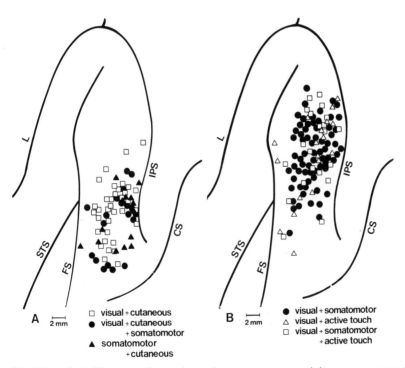

A   2 mm   □ visual + cutaneous
  ● visual + cutaneous
    + somatomotor
  ▲ somatomotor
    + cutaneous

B   2 mm   ● visual + somatomotor
  △ visual + active touch
  □ visual + somatomotor
    + active touch

Fig. 75 A, B. A The recording points where cutaneous activity was represented together with visual, or somatomotor, or visual and somatomotor activation. **B** The recording points where visual activity was represented together with somatomotor functions, active touch, or both of them. (Hyvärinen 1981c)

Table 9. Lateral distribution of combined responses (see Fig. 75).
(Hyvärinen 1981c)
Group A = visual + cutaneous, somatomotor + cutaneous,
visual + cutaneous + somatomotor.
Group B = visual + somatomotor, visual + active touch, visual
+ somatomotor + active touch

	Mean y, mm	S.D.	Difference
Group A	21.6	4.3	8.5[a]
Group B	13.1	5.1	

[a]$P < 0.001$, Student's t-test

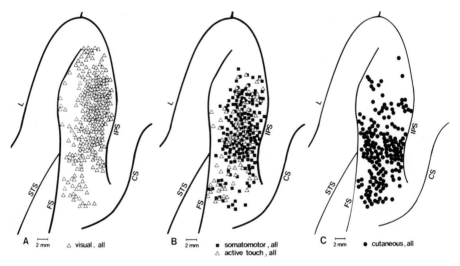

Fig. 76 A-C. Distributions of all visual points (**A**), all somatomotor and active touch
points (**B**), and all cutaneous points (**C**). At some points in each map responses were
also evoked by other functions than the one indicated in the legend. (Hyvärinen 1981c)

    Above, the modality-pure responses and combined responses were
presented separately, and they showed great regional differentiation. However, when different types of sensory or motor responses were analyzed
regardless of whether they occurred alone or in combination with other
response types, somewhat different results were obtained (Table 7,
group 3).
    Figure 76A indicates that the distribution of all the points with visual
responses, alone or in combination, extended more laterally than the distribution of only visual points (compare with Fig. 72A) and covered area
7b, too. However, these points did not cover the posterior part of 7a
where the kinaesthetic responses occurred. Figure 76B shows all the

points with responses related to somatic movement and active touch. These points covered approximately a similar area as the corresponding pure responses presented in Fig. 72B. Figure 76C shows all the points with cutaneous responses. As a whole this group was located more laterally than the visual points; they covered by and large area 7b, as did the cutaneous points with modality-pure responses presented in Fig. 72C.

C. Somatotopy in Area 7

Motor responses indicated a clear somatotopy in area 7 as motor responses related to the eyes, arm or mouth had significantly different distributions (Fig. 77A). Oculomotor responses were located most medially and responses related to the mouth most laterally. Some degree of somatotopy was also evident in cutaneous responses from different body parts (Fig. 78). The cutaneous responses from the snout were located most laterally, whereas the ones from other parts of the head were located more medially (Fig. 78A, Table 10). The distributions of the hand and head differed significantly from the rest except each other, whereas the arm, trunk and leg formed one group inside which no significant differences were observed.

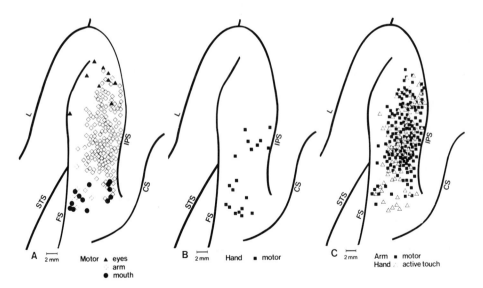

Fig. 77 A-C. A Motor responses related to movements of the eyes, arm and mouth. **B** Motor responses related to movements of the hand. **C** Responses related to motor movements of the arm and active touch with the hand. (Hyvärinen 1981c)

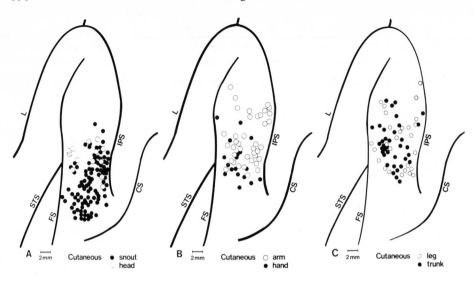

Fig. 78 A-C. Somatotopic arrangement of the cutaneous representation. **A** Cutaneous responses from the snout and from the rest of the head. **B** Cutaneous responses from the hand and arm. **C** Cutaneous responses from the trunk, and leg and foot. (Hyvärinen 1981c)

Table 10. Lateral distribution of points with cutaneous responses from various body parts (Fig. 78). (Hyvärinen 1981c)

Body part	n	Mean y, mm	S.D.	Differences and their significances					
				Snout	Hand	Head	Arm	Trunk	Leg and foot
(1) Snout	94	22.8	4.1	0	3.9^c	4.6^c	7.8^c	6.9^c	7.5^c
(2) Hand	18	18.9	3.7	3.9^c	0	0.6	3.9^b	3.0^b	3.6^b
(3) Head	13	18.2	2.5	4.6^c	0.6	0	3.2^b	2.3^a	2.9^a
(4) Arm	39	15.0	4.2	7.8^c	3.9^b	3.2^b	0	0.9	0.3
(5) Trunk	36	15.9	3.8	6.9^c	3.0^b	2.3^a	0.9	0	0.6
(6) Leg and foot	27	15.3	4.8	7.5^c	3.6^b	2.9^a	0.3	0.6	0

$^a p < 0.05$, $^b p < 0.01$, $^c p < 0.001$, Student's t-test

D. Functional Differentiation

The above results show that there are several *functionally different zones* in the exposed part of the medial parietal gyrus, and that these areas overlap to a great extent. *Some somatotopy is present* in this area, too, although it cannot be compared with that present in SI or SII. In the exposed part of the medial parietal gyrus the purely visual functions were concentrated in the medial part of area 7. However, in the studies described in Chapter VIII we have found inside the intraparietal sulcus single neurones related to visual and oculomotor mechanisms more laterally than in this mapping study (Hyvärinen and Poranen 1974, Leinonen et al. 1979). On the other hand, when neuronal activity related to visual stimuli was studied without regard to somatic mechanisms, which may also activate the cells, the visual representation was found to cover the whole medio-lateral extent of area 7.

The differences observed between areas 7a (PG) and 7b (PF) also fit with what is known of the anatomical connections of these areas (Table 2). Only the most medial part of area 7a, where the oculomotor responses were found (Fig. 72A), projects to the superior colliculus (Kuypers and Lawrence 1967), and only 7a is connected reciprocally with the eye movement area of the frontal lobe (Pandya and Kuypers 1969). Projections from areas 18 and 19 arrive in the medial part of area 7 (Pandya and Kuypers 1969, Stanton et al. 1977) whereas area 2 and SII send afferents only to 7b (Pandya and Kuypers 1969, Vogt and Pandya 1978, Stanton et al. 1977).

In my laboratory Leinonen and Nyman (1979) found an associative area related to the face close to the lateral tip of the intraparietal sulcus. The maps presented above show that the part of area 7b related to the face is actually quite large, covering the most lateral part of area 7b. As is typical of area 7 this part also contains visual mechanisms differing in this respect from the anterior parts of the parietal lobe (areas 3, 1, 2, and 5). It is interesting that the anatomical arrangement of the postcentral and medial parietal gyri is such that area 7 meets the primary somatosensory cortex at the lateral level of the representation of the mouth. The somatosensory cortex does not contain visual mechanisms, and area 7 does not contain a representation of the inside of the mouth. Evolution of this structure is logical because there is no use of visual interaction inside the mouth where somatosensory and gustatory discriminations prevail. On the other hand, vision is very useful in guiding arm and hand

Table 11. Some cortical mechanisms of functional coupling. (Hyvärinen 1981a)

Body part	Functional coupling	Cortical area
Arm-hand	Somatic-visual	Area 7
Head	Somatic-auditory	Area Tpt
Inside of mouth	Somatic (-chemical?)	SI

movements and even movements of the face and lips (but not those of the tongue), functions that are all represented in area 7.

Three different cortical mechanisms of this type of functional coupling are listed in Table 11 (Hyvärinen 1981a). These are only illustrative examples of functional coupling in the parietal cortex in the monkey; to some extent the same functions also have other representations. In area 7 the movements of the hand-arm system are guided by the use of vision during reaching for objects of interest. In area Tpt auditory activation and localization mechanisms are coupled with turning of the head, but for active sensory discrimination of objects inside the mouth senses other than somaesthetic (and chemical) are not useful. Therefore, the inside of the mouth is not represented in the posterior parietal association area but in SI in the anterior parietal lobe.

The somatomotor region was located significantly more anteriorly than the kinaesthetic region (Table 8). It is possible that in the somatomotor region sensory afferentation from joints and muscles contributes to the activity during movements, thus giving feedback information about the movements. In this sense the somatomotor neurones may be influenced by a similar sensory element as the active touch neurones, which are activated through the skin during voluntary movements. The somatomotor neurones, however, receive their sensory input from joints and muscles. The joint and muscle information for the guiding or somatic movements may reach the somatomotor part of area 7 through the kinaesthetic region, but other routes are also possible, e.g. via the pulvinar (Burton and Jones 1976, Pearson et al. 1978) or area 5 (Jones and Powell 1969, 1970b). It is also possible that area PGa inside the superior temporal sulcus (Selzer and Pandya 1978) contributes connections to the kinaesthetic area. Connections to the kinaesthetic area may also arrive from SI directly or via area 5 (Pandya and Kuypers 1969, Jones and Powell 1970b, Mesulam et al. 1977, Vogt and Pandya 1978). Muscle and joint activation has not been described in SII (Whitsel et al. 1969, Robinson and Burton 1980c) which therefore is not a likely projection route for these responses. On the other hand, SII probably sends cutaneous input to area 7b since many of the properties of the neurones in 7b activated through the skin resemble those of the neurones in SII (Whitsel et al. 1969, Robinson and Burton 1980c).

My mapping study did not cover that part of area 7b which extends into the Sylvian fissure as did the studies of Robinson and Burton (1980b, c). They also found somatotopic arrangement in area 7b outside SII. The borders of SII in the monkey were previously somewhat uncertain, because of the difficulties encountered by Woolsey and Fairman (1946), who stated that somatic area II responses are obtainable in the monkey only under quite light anaesthesia. These borders have been largely clarified by Friedman et al. (1980) and Robinson and Burton (1980a, b, c). In addition to the somatic body representation in SII these authors have obtained results indicating that various body parts are represented in area 7b (Robinson and Burton 1980b). The maps presented above, together with the results of Robinson and Burton (1980b), Leinonen et al. (1979), and Leinonen and Hyvärinen (1980), suggest that another complete body representation may exist in the lateral parts of area 7 and outside SII. After extracting the information of all these maps, one gains the impression that in the most lateral part of area 7 the face is represented anteriorly, the upper limbs behind the face; further in the upper lip of the Sylvian fissure lie the representations of the whole body and the lower limbs.

The existence of a complete body representation in area 7b together with visual mechanisms may aid area 7 in its task to orientate different body parts in the immediate extrapersonal visual space (see Chap. XII). However, so far no single study has covered all of area 7b; such a study could indicate whether this interpretation of the somatic representation in area 7b is correct. Until then it must be considered as tentative.

XI. Modification of Area 7 and Functional Blindness After Visual Deprivation

An old and interesting problem concerns the question whether the visual system of the brain is innately, without visual experience, prepared to handle visual information. This problem was already formulated in 1690 by John Locke as the *question of Molineux*.

This question is quoted in Locke's *An Essay Concerning Human Understanding* (1690) as follows: "... I shall here insert a problem of that very ingenious and studious promotor of real knowledge, the learned and worthy Mr. Molineux, which he was pleased to send me in a letter some months since; and it is this: 'Suppose a man *born* blind, and now adult, and tought by his *touch* to distinguish between a cube and a sphere of the same metal, and nighly of the same bigness, so as to tell, when he felt one and the other, which is the cube, which the sphere. Suppose then the cube and sphere placed on a table, and the blind man be made to see: *quaere*, whether *by his sight, before he touched them*, he could now distinguish and tell which is the globe, which the cube?' To which the acute and judicious proposer answers: 'Not. For, though he has obtained the experience of how a globe, how a cube affects his touch, yet he has not yet obtained the experience, that what affects his touch so or so, must affect his sight so or so; or that a protuberant angle in the cube, that pressed his hand unequally, shall appear to his eye as it does in the cube.' — I agree with this thinking gentleman, whom I am proud to call my friend, in his answer to this problem; and am of opinion that the blind man, at first sight, would not be able with certainty to say which was the globe, which the cube, whilst he only saw them; though he could unerringly name them by his touch, and certainly distinguish them by the difference of their figures felt."

Experiments of the kind mentioned by Mr. Molineux have occurred occasionally; less than 100 cases have been documented in the literature. The weight of the evidence overwhelmingly supports the conclusion of Mr. Molineux, i.e., the visual system of the brain is not innately, without visual experience, capable of extracting the meaning of visual signals.

Von Senden (1932) studied 66 cases mentioned in the literature during this millenium (between years 1020 and 1931) where a person was documented as blind at early age and operated on in later years so as to gain sight. He concluded that persons born blind and later operated on had not acquired an understanding of spatial relations which above all requires visual perception, and that this understanding was not gained without difficulty after the operation; the patients had to work hard for it.

This, furthermore, led to dramatic change and often to strong conflicts in their lives. To reduce the difficulties these people tended, whenever possible, to make use of their capacity for tactile experience. Moreover, prior to operation these people were unaware and unable to conceive that they were observed by other people during their work and daily activities. After the operation they often became acutely aware of such observation which made them try to please others. This could further aggravate their mental handicap in relation to seeing people.

Gregory and Wallace (1963) presented an illustrative case: A man with slight residual vision in childhood developing later to complete blindness had a successful corneal operation at the age of 52. This patient was able to learn to make elementary use of his sight although he never gained enough vision to recognize facial expressions. Soon after the operation he became depressive and died a few years later.

Valvo (1971) wrote an interesting report on four blind patients operated on successfully by Strampelli in Rome, who fitted osteo-odonto-keratoprostheses made from a tooth to their eyes. One of these patients was a keen observer who kept a detailed diary during the recovery of his vision. His notes reflect the difficulties encountered by a patient when shifting from tactile to visual observation, although these difficulties were eased by the fact that he was already 16 years old at the time of losing his sight. During personal interview this patient impressed me with two recollections from the period after the operation. A couple of months after regaining sight his wife asked why he was constantly watching his hands and playing with them. Although he was not at first fully conscious of this activity his conclusion after thinking of it was that he wanted to learn to use the hands under visual guidance and this required a lot of training and learning. On the other hand while exploring objects he was occasionally confused of whether he should "actually touch objects with his eye or see them with his hand". This confusion reflects the slow process of shifting from manual to visual exploration. However, when recognizing objects, people and animals by sight he experienced strong emotional feelings which he compared with the feelings of a small child who is learning new things (Strampelli et al. 1969).

A particularly well-documented case was reported by Ackroyd et al. (1974). Their female patient blind from the age of 3 received corneal graft at the age of 21; she was bright and motivated for learning to use her vision. Although she learned to detect, locate and follow the movement of conspicuous objects, to avoid large obstacles and to make some discriminations of brightness, she remained unable to recognize objects by their visual form. She considered the operation to be a failure, became depressive, and then resumed her life as a blind person.

Studies of congenitally blind children emphasize the importance of *early eye contact* with the mother for a stable mother-infant relationship and the *use of the hand under visual guidance* for understanding of spatial relations and active movements (Fraiberg and Freedman 1964). Blind children without these experiences face a great risk of becoming autistic (Fraiberg 1977, Warren 1977). Also in a congenitally blind child the gaining of sight may lead to emotional difficulties if the transition is inadequately guided (Segal and Stone 1961).

Experimental studies on animals indicate modifiability of visual functions at an early age. For instance Held and Bauer (1967) described infant monkeys who lacked experience of visual guidance of arm movements and who were unable to use their hands appropriately for a few weeks after being allowed to use visual guidance. Their reactions to the arm when first seeing it were like towards a foreign object that they started to become familiar with. Thus it appears that experience of eye-hand coordination is necessary for the perfection of this function. Such experience may affect the neurones in area 7 that are related to reaching with the arm, but no recordings have yet been performed in monkeys reared this way.

Effects of *binocular deprivation* were studied at the end of the last century by Hans Berger, who later was to become the renowned discoverer of the electroencephalogram. He conducted experiments on kittens and puppies, studying anatomical effects of binocular lid closure (Berger 1900). Sixty years later Riesen showed in a series of studies on cats, monkeys and chimpanzees (Riesen 1958, 1961a, 1966, Riesen and Aarons 1959, Riesen et al. 1964) that the visual acuity of animals reared in darkness was low although it improved with time. However, such visually guided behaviours as visual placing, avoidance of the apparent depth on the visual cliff (transparent bridge) and reaching with hand were very slow to develop in monkeys after such rearing. Riesen emphasized that function of neural pathways was necessary for their development (Riesen 1961b).

In the 1960's Wiesel and Hubel (1963, 1965) studied the effects of *monocular deprivation* of form vision in kittens. They showed that during an early sensitive period such deprivation produces a profound reduction in the effectiveness of the deprived eye to influence cortical neurones. Similar results were also obtained in young monkeys (Baker et al. 1974, Blakemore et al. 1978, von Noorden and Crawford 1978). However, the results of Wiesel and Hubel (1965) on binocular deprivation were more complex, since this type of deprivation led to behavioural blindness in kittens although after opening of the eyes over half of the cells in the visual cortex were activated by visual stimuli. They suggested that the deprivation effect observed in the visual cortex after monocular closure depended on *competition between the inputs from the two eyes* rather than disuse as such, and that the blindness observed after binocular closure was caused by impairment of levels central to area 17 in the visual system.

These results suggested a hypothesis that at an early age competition might occur between the different inputs to the association cortices. Since area 7 receives input from two sensory systems, vision and somaesthesia, the disuse of one of them (in these studies vision) could lead to a decrease in its neural representation. Such neural changes in association areas of the brain could explain the persistent effects that deprivation has on the visual behaviour of animals and humans observed after correction of abnormalities that blocked vision at an early age. For investigation of this problem we conducted a series of studies on young monkeys (Hyvärinen et al. 1978a, Hyvärinen and Hyvärinen 1979, Hyvärinen et al. 1981a,b,c).

A. Visual Deprivation

We raised five laboratory-born stumptail monkeys with bilateral lid closure performed shortly after birth. These baby monkeys were taken care of by their mothers in individual cages. A play pen with a small entrance was arranged beside the mother's cage and it was actively used by normal and deprived baby monkeys (Hyvärinen et al. 1978c). The lids of these monkeys were opened at the ages of 7 to 11 months under pentobarbital anaesthesia during the operation in which the recording apparatus was fixed on the skull.

After opening the lids the optic media were clear, the ocular fundi appeared normal and pupillary light reflexes were present. The refractions ranged between +1.5 and +6.0 diopters and did not differ significantly from those observed in normal monkeys. Spontaneous irregular nystagmus developed in a few days after the opening of the eyes but it subsided within a couple of weeks after which optokinetic nystagmus could be elicited in all animals inside a rotating drum covered with vertical black and white stripes of 15°. In their own familiar cages using somaesthetic cues the deprived monkeys learned to move almost equally well as their normal peers, but in unfamiliar surroundings the mobility of the deprived infants was markedly impaired.

Behaviourally all the deprived monkeys were blind (Fig. 79). They bumped into obstacles, fell from tables and were unable to reach for food with the hand or mouth under visual guidance. Visual placing reactions were not developed by them although tactile placing was normal, and they never showed teeth-chattering response to threatening facial expressions although this innate response typical of macaque monkeys (described, e.g., by Bertrand 1969) was readily evoked by threatening voices. In the colony

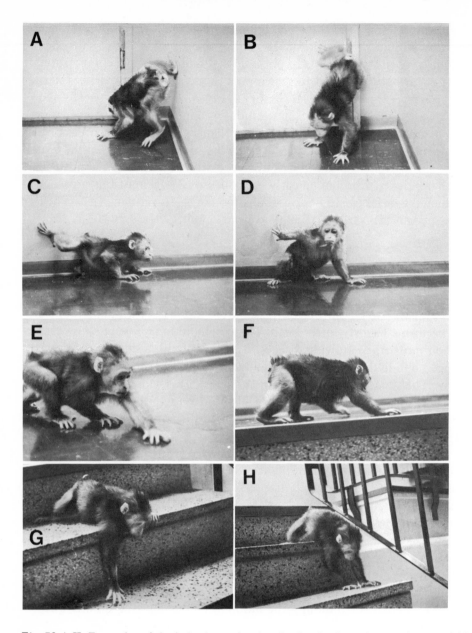

Fig. 79 A-H. Examples of the behaviour of a visually deprived monkey at the age of 15 months 4.5 months after the eye opening. The monkey still lacks visually orienting behaviour as is seen in this series of photographs taken when the monkey was put in an unfamiliar room. **A** When put in a strange place the monkey starts to feel the limits of the space raising its foot on the wall and keeping the hands safely on the floor. **B** It places its both feet higher and higher to find out whether there is any upper limit to the surface or something to climb on. **C** Finding nothing in the upward direction the monkey starts to move alongside the wall to which it keeps contact with the foot feeling the floor simultaneously with the hands. **D** It is somewhat

they were uninfluenced by visual communication of the other monkeys although auditory signals and somatic contacts triggered normal behavioural patterns. Because of harassment by other monkeys caused by lack of comprehension of visual signals these monkeys could not be kept with monkeys other than their mothers for any length of time.

The visual behaviour of three deprived monkeys was observed during one month after opening of the eyes. During this time no recovery of visual behaviour was observed in them; their behaviour remained as it was immediately after opening of the eyes. In addition one monkey has been observed for 6 months and one for 3 years after opening of the eyes. In these two monkeys some recovery of visual behaviour has occurred, although they are still considerably visually handicapped. They are now able to avoid obstacles in their path through use of vision, and they sometimes look at objects in their hands at a close distance. Their mobility is, however, slow and cautious in comparison with the normal animals, and they still do not react to facial expressions.

The poor visual behaviour of these monkeys resembles the behaviour of humans whose ocular abnormalities have been corrected late in life as described above. The visual orientation behaviour of these monkeys appeared similar to that described in humans gaining sight, and it also resembled the descriptions of the behaviour of cats shortly after the end of dark-rearing. However, the monkeys did not gain normal visual skills after opening of the eyes as has been described to happen with cats. Cats have been reported to recover full vision after dark-rearing within a couple of months (Baxter 1966, van Hof-van Duin 1976, Timney et al. 1978). Although the visual acuity of monkeys also recovers after dark-rearing of 3 or 6 months when tested with a forced-choice preferential looking test, the behavioural responsiveness to visual stimuli remains impaired (Regal et al. 1976). In our monkeys the recovery of vision after binocular lid closure of 7 to 11 months was minimal during the first months after the opening of the eyes. In this respect monkeys, being strongly visual animals, may be closer to man than to the cat.

◄ afraid of leaving the wall that gives some security. E Finally it leaves the wall and moves carefully with the limbs slightly flexed and one arm extended in front to feel possible obstacles. F When the monkey encounters an edge, it changes direction. G When lifted one step down the stairs it begins to feel downwards to get tactile information of the surroundings. H Finally it puts its both hands on the lower step before descending carefully. (Hyvärinen et al. 1981c)

B. Deprivation Effects on the Visual Pathways

We studied visual activity in areas 17 and 19 in the deprived animals and normal controls using multiple unit recording technique. Because of the poor visual behaviour of the deprived animals they could not be trained to fixate. Therefore untrained animals free to move their eyes were studied. The recording methods were similar to those used in the mapping studies presented in Chapter X. Various kinds of complex visual stimuli consisting of black-and-white and coloured patterns, flashing lights, food and other objects were shown to the animals. When no response was obtained to such stimuli, discharge was often elicited by uncovering the animal's eyes. When during an electrode penetration stable spontaneous activity was recorded that was not consistently associated with any type of stimulation nor with any maneouvers performed by the animal, the recording site was classified as only spontaneously active.

Strong responses to visual stimulation were recorded from the deprived animals during penetrations made in the visual cortex in area 17 and in area 19 along the inferior parietal gyrus. As regards these results the deprived monkeys did not differ from normal animals, but it would require single-cell technique in anaesthetized and paralyzed animals to detect possible differences in the receptive fields, properties of disparity, etc. between cortical cells in normal and deprived animals. A slight change in the ocular dominance distribution was observed in area 19 in favour of the contralateral eye as a consequence of deprivation (Fig. 80) (Hyvärinen et al. 1981c).

In another study (Hyvärinen et al. 1981a) somatic responses were observed in area 19 (Fig. 81A) of deprived animals (Table 12). All of the 131 recording sites studied in area 19 of normal animals responded to visual stimuli, but in the deprived animals only 43 recording sites of the 110 studied responded to such stimuli (Fig. 81B), whereas 20 of them were *activated during the monkey's active manipulation of objects* (Fig. 81C). As indicated in Fig. 81A, somatic responses were recorded in all parts of the area of study. The points with visual responses and the ones only spontaneously active were also evenly distributed. Manipulation of objects never elicited cellular responses in area 19 of normal animals in which this response type is common in the lateral part of area 7. In the deprived animals 47 recording sites were spontaneously active but could not be activated with sensory stimuli and were not related to any obvious motor behaviour. No recording sites were activated by both visual and somatic mechanisms, suggesting that the visual and somatic neurones were segregated.

In this study the adjacent part of area 17 was also studied in 91 electrode penetrations. All these recording sites were activated by visual stim-

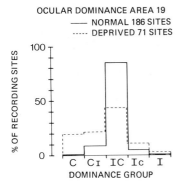

OCULAR DOMINANCE AREA 19
—— NORMAL 186 SITES
····· DEPRIVED 71 SITES

I= IPSILATERAL, C = CONTRALATERAL

Fig. 80. Ocular dominance distributions in area 19 of the visual cortex revealed by recording of multineuronal impulse activity at 186 sites in three normal monkeys and at 71 sites in one deprived monkey. Dominance groups as in Fig. 73. (Hyvärinen et al. 1981c)

Table 12. Activating mechanisms in area 19 of three normal and three binocularly deprived monkeys. Numbers of multiple unit recording points made in the area hatched in Fig. 81A. (Hyvärinen et al. 1981a)

	Normal		Deprived	
	N	%	N	%
Visual	131	100	43	39
Somatic	0	0	20	18
Only spontaneous	0	0	47	43
Total	131	100	110	100

uli. However, neurones at 16 recording sites located close to the 17−18 border were in addition to visual responsiveness weakly activated during manipulation of objects, particularly food or other objects that the monkey was interested in. Area 18 was not studied.

Monocular deprivation in kittens and monkeys leads to deterioration of the connections from the deprived eye to neurones in area 17 which is consistent with a decrease in the visual acuity of the deprived eye in such animals (Wiesel and Hubel 1963, 1965, Baker et al. 1974, Blakemore et al. 1978, von Noorden and Crawford 1978) and in man (Awaya et al. 1973). The cortical deprivation as a sequel of early monocular closure is best explained as a result of competition between the inputs from the two eyes: one input deteriorates if and only if the other input is active. Thus binocular deprivation does not affect the connections from each eye to the cortex nearly as severely as monocular deprivation. However, some changes are produced in the visual pathways by binocular deprivation, too. Such chang-

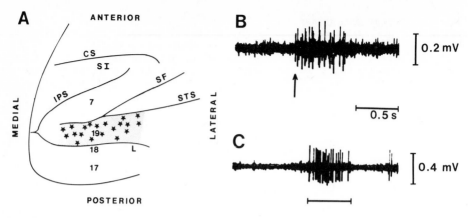

Fig. 81 A-C. A The region of recording (*hatched*) in area 19 depicted on the right hemisphere. Abbreviations: *CS* central sulcus; *SF* Sylvian fissure; *IPS* intraparietal sulcus; *STS* superior temporal sulcus; *L* lunate sulcus; *SI* primary somatosensory area. Brodmann's areas 7, 19, 18 and 17 are also marked on the figure. The points where somatic responses were found are indicated with *starlets*. **B** Neuronal response to opening of the eyes (*arrow*) in area 19 of a binocularly deprived monkey. **C** Neuronal response observed when the monkey was manipulating the experimenter's hand in search for a raisin. The duration of manipulation is indicated with a *bar* under the record. Similar responses were obtained with eyes open and eyes covered. (Hyvärinen et al. 1981a)

es are known to occur in the thalamus (Headon and Powell 1978) and in the visual cortex (Wiesel and Hubel 1965, Crawford et al. 1975), although they are less marked than after monocular deprivation. Wiesel and Hubel (1965) observed in binocularly deprived kittens that 41% of the neurones recorded in area 17 had normal receptive fields, 32% responded to visual stimuli but had abnormal receptive fields and 27% could not be driven with visual stimulation. Moreover, others (Pettigrew 1974, Blakemore and van Sluyters 1975, Buisseret and Imbert 1976, Leventhal and Hirsch 1980) have later shown that the receptive field specificity typical of normal visual cortex is lacking or greatly reduced in binocularly deprived cats. These data indicate that although the visual cortical areas are responsive to visual stimulation, a change in the visual properties of the neurones may contribute to the poor visual behaviour of binocularly deprived animals. However, as explained below, a much greater change in the visual properties of the neurones is observed in the parietal association cortex.

The somatic activation found in area 19 indicates that binocular deprivation profoundly alters the function of this area. Normally area 19 appears entirely related to visual functions (Zeki 1978, van Essen 1979) although the responses may be independent of passive visual stimulation. As Fisher and Noth (1981) have recently shown the activity of prelunate neurones in normal monkeys is related to selection of targets for foveation. The fact that all recording sites in our normal monkeys responded to visual stimulation may thus indicate that the stimuli were of interest to the monkeys who therefore foveated them. In the deprived animals, however, fewer neurone groups responded to such stimuli, but many were related to active sensing with the hand. This form of activity could be characterized as "manual foveation", i.e., direction of somatic exploration towards interesting targets. In the deprived animals mechanisms related to such "manual foveation" may have replaced some of the activity normally related to selection of visual targets for foveation.

The influence of stimuli of other modalities on the responses of the visual cortex of the cat has previously been emphasized (Jung et al. 1963). Although electric shocks on the skin (Horn 1963) and painful pinpricks (Murata et al 1965) influence the activity in the cat's visual cortex, it has not been found to respond to the stimulation of mechanoreceptors or joints. However, area 19 may receive synaptic influences from other structures that could mediate the somatic responses. Anatomical studies show that the thalamic posterior, lateral posterior, and pulvinar nuclei could act as sources of such input (Graybiel 1970, Heath and Jones 1970, Rosenquist et al. 1974, Maciewicz 1975, Benevento and Rezak 1976, Trojanowski and Jacobson 1977). In the deprived condition the lack of activation of the visual pathways to area 19 may lead to a *compensatory increase in the synaptic efficiency in the pathways mediating somatic inputs*. It remains an open question whether area 19 in the visually deprived can make any use of the information it receives from active touch during pattern perception tasks.

C. Deprivation Effects on Area 7

Neuronal multiple unit activity was recorded in area 7 of four binocularly deprived and three normal young monkeys using methods similar to those in the mapping studies described in Chapter X (Hyvärinen et al. 1981c). Recordings extended to both areas 7a and 7b. These recordings indicated a *great reduction in the proportion of cell groups activated by visual stimulation as a consequence of binocular deprivation* (Table 13). Moreover, the few visually drivable cell groups found gave only weak responses to visual stimulation. The proportion of purely visual neurone groups that constituted 24% of the sample in the normal animals was reduced to 2% in the deprived animals, and the reduction in the proportion of cell groups activated by both visual and somatic mechanisms was

Table 13. Occurrence of different types of activity in intracortical multiple unit recordings made in area 7 at 399 recording sites in 2 young normal monkeys and at 426 recording sites in 4 binocularly deprived monkeys after opening of the eyes. (Hyvärinen et al. 1981c)

Type of activity	Normal monkeys		Deprived monkeys		Change as % of normal proportional representation
	Number	Proportional representation	Number	Proportional representation	
	N	%	N	%	%
Visual	97	24	11	2	– 92
Visual and somatic	121	30	4	1	– 97
Active somatic movement	94	24	220	52	+117
Passive somatosensory	76	19	93	29	+ 53
Only spontaneous	11	3	68	16	+433
Total	399	100	426	100	

even more drastic (from 30% to 1%). In the deprived animals the representation of all visual mechanisms combined was only 6% of that normally observed in area 7. The proportion of neurone groups functioning during active somatic movement was increased by 117% and the proportion of those activated by passive somaesthetic stimulation by 53%. In the deprived animals many of the cell groups that functioned during active movement were activated during minor, delicate finger movements when the monkey was manipulating an object and its hands encountered small differences in texture, edges, etc., although no other form of stimulation was effective. In both groups of animals the medio-lateral extent of area 7 was covered by recordings roughly equally; thus these differences did not result from unequal sampling from different parts of area 7. The number of cell groups that could not be activated by external stimulation and were not active during movements performed by the monkey, i.e., cell groups that were only spontaneously active, had increased fourfold in the deprived animals. These changes are indicated graphically in Fig. 82.

In one monkey we recorded multiple unit activity in area 7 after 2 years had elapsed from the opening of the eyes. As Table 14 indicates, some recovery of visual input to area 7 had taken place. However, in agreement with the poor visual behaviour of the animal the percentage of visually activated cell groups still remained low, being about one quarter of the visual representation observed in normal animals (Hyvärinen et al. 1981b).

The lack of visual orientation behaviour which we observed in our monkeys and which has been observed in monkeys and cats after dark-rearing (Riesen 1958, Baxter 1966, Regal et al. 1976) is not in harmony with the visual responses observed in areas 17 and 19 but fits with the nearly total lack of visual function in association area 7. Moreover, Regal et al. showed that after dark-rearing monkeys' general visual behaviour may be altered more than visual acuity per se. Our monkeys, too, suffered from

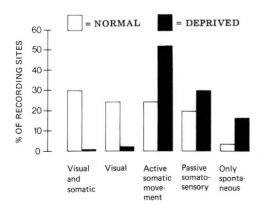

Fig. 82. The representation of various types of activities in area 7 of the parietal association cortex in normal and visually deprived young monkeys. 399 recording sites were studied in normal and 426 in deprived monkeys. (Hyvärinen et al. 1981c)

Table 14. Recovery of functions in area 7 two years after opening of the eyes in one monkey. The data on normal monkeys are from Table 13. (Hyvärinen et al. 1981b)

Type of activity	Normal monkeys		One deprived monkey		Same monkey two years later	
	N	%	N	%	N	%
Visual or visual + somatic	218	54	5	3	13	13
Somatic (active or passive)	170	43	140	88	86	86
Only spontaneous	11	3	14	9	1	1
Total	399	100	159	100	100	100

defects in general visual behaviour and orientation, functions that may be related to associative cortices, whereas visual acuity is probably related to functional properties of the primary visual pathways.

The most marked effect of deprivation on responses in area 7 was the *almost complete absence of cell groups activated both by vision and somatic mechanisms,* a combination that is common in normal animals in area 7 (see Chap. VIII.C), and in the posterior periarcuate region of the frontal association cortex (Rizzolatti et al. 1981). This combination illustrates the role of the association cortex in the integration of visual and somatic inputs needed for orientation in the surroundings and in the immediate extra-personal space. Binocular deprivation had its greatest effect on the visual aspect of such integrative functions. The ability of cells in the visual cortex to respond to visual stimuli was not abolished as a consequence of binocular deprivation, but the further integration of viusal information with other modalities and motor actions was. Moreover, the representation of exploratory use of the hand was increased. Recovery from early deprivation effect appears to be very poor in monkeys. Even after 2 years of recovery, area 7 of one monkey contained only one quarter of the normal visual representation. This finding helps to explain why it is so difficult for persons deprived of vision at an early age to achieve good functional vision later in life.

These findings could be explained by *extending the principle of competition between two inputs introduced by Wiesel and Hubel (1965) to concern the parietal association cortex.* Here the visual and somatic inputs are competing; the reduction in the activity of the visual pathway results in lack of visual functions in this area and probably also in other visual association areas in the temporal and frontal lobes.

Since binocular deprivation also produced changes in the visual areas, it is possible that two different mechanisms contribute to the behavioural effects: the cross-modal competition effect and the accumulation of lesser deprivation effects at various levels of the visual pathways and areas. However, the visual representation in area 7 was more susceptible to binocular deprivation than the visual response in areas 17 and 19 suggesting that the cross-modal competition effect may be greater than the disuse effect on the visual system proper. The somatic responses found in area 19 in deprived animals could be taken as another example of cross-modal competition. Due to lack of use of the normally dominant visual input the dormant, only subliminally activated somatic inputs to this region gain efficiency as a consequence of visual deprivation.

The large increase in the proportion of cell groups only spontaneously active in areas 19 and 7 indicates that all visual input was not replaced by somatic function. The lack of driving in these neurones may be related to the lessened overall activity of these animals and their slow mobility in unfamiliar surroundings based on lack of understanding of visual space. The increase in the number of unresponsive neurones indicates a general decrease in the effciency of association mechanism caused by early blindness which in primates cannot be compensated by other forms of experience.

In the light of this study, the lasting symptoms of early visual deprivation in humans and monkeys arise from the changes that this condition brings about in the associative systems of the brain leading to a decrease in the efficiency of neural transmission along pathways that mediate visual influences in the associative systems. This finding *stresses the importance of early detection and correction of visual defects to avoid permanent changes in the neural mechanisms.*

In retrospect it was rewarding to learn that at the turn of the century the great neuroscientist Santiago Ramón y Cajal was far-sighted enough to predict our results on the association cortex when he wrote in his *Recollections of My Life* (1937): "It seems probable that the singular idiosyncracies of certain brains are due not merely to chance augmentation or to a perfection by use of certain cells and pathways, but also to focal failures of neuronal growth, as a result of which particular association systems may be extraordinarily weak or even absent."

XII. Functional Role of Parietal Cortex

In this concluding chapter I shall attempt to draw some conclusions as to the functional role of the parietal cortex. It has been established that the anterior part of the parietal lobe, the postcentral gyrus, is involved in somaesthesis and somatic guidance of movement. Its functional role thus appears clear although the differences found between various cytoarchitectural areas require further thought. Area 5 being a somatosensory associative area belongs to the posterior parietal association cortex and is, regarding both localization and function, in an intermediate position between the primary somatosensory cortex and area 7 in the posterior parietal lobe.

A discussion of the functional role of the posterior parietal cortex is not easy (Hyvärinen 1982). This cortical region is involved in a number of functions, described in Chapters V to IX, whose reciprocal action should be understood in order to form a coherent view of the function of the whole area. To emphasize any single one of these functions at the expense of the others would result in a biased view of the functional significance of posterior parietal cortex. On the other hand, although the functions of the posterior parietal cortex are various, its role is not as extensive as the role of the whole brain. The temptation to ascribe vast functions to the posterior parietal cortex is appreciable since, the functions of this part of cortex include sensory, motor and behavioural activities. However, it is obvious that the posterior parietal cortex hardly performs independently any of its functions; it makes extensive use of the multiplicity of incoming and outgoing connections with other regions of the brain.

A. Somatosensory Cortex

The primary somatosensory cortex is organized strictly *somatotopically*. Its neurones have rather small receptive fields that facilitate accurate localization of stimuli, and they respond briskly, indicating precisely the *timing* of sensory events. Moreover, in SI different submodalities are segregated in columns giving a *localizational basis for stimulus quality*.

This arrangement is of the type required for somaesthetic guidance of motor performance. Such a system is able to give accurate sensory feedback information during motor actions. This feedback information may concern the location (on the body), timing and quality of external stimuli and proprioceptive events originating from subcutaneous receptors, muscles and joints.

However, these functional properties are not well suited for more general somatic orientation, personal awareness and shifts of attention from one body part to another. These functions are more likely served by the second somatosensory area where the neurones have large receptive fields and respond to stimuli with less accurate timing, and where some neurones respond only to nociceptive stimuli (Robinson and Burton 1980c). For more general orientation of the body in the surrounding space these somatic mechanisms combine with visual activation in area 7b. Thus SII may contribute to a global *autotopognostic body scheme*, whereas the role of SI in a body scheme may be in focussing to smaller details.

The various postcentral areas have specialized in different kinds of somatic tasks. Area 3a receives input from muscles (Phillips et al. 1971), areas 3b and 1 from the skin, area 2 from the skin and joints, and area 5 mainly from joints and also from the skin and muscles. None of these areas contains mechanisms limited to a single submodality, however. They thus all participate in various somatosensory activities but may each have dominant roles in different functions. The muscle input to area 3a is suitably located between the sensory and motor areas to serve sensory feedback from active muscles. On the other hand, the strong cutaneous input to areas 3b and 1 mediates signals from the surfaces of external objects that the skin encounters during movements. The separate columns specific for slowly or rapidly adapting receptor input (Sur et al. 1981) may convey additional information about the quality of the cutaneous touch. Area 2 represents a transitional zone between the anterior postcentral gyrus and the posterior parietal area 5. Area 2 thus contains a representation of joints and convergence of inputs from different submodalities. Area 5 again is an associative somaesthetic area that combines inputs from the joints with input from various submodalities to represent different somatosensory patterns arising during movement. The joints represented in area 5 are typically the large joints of the limbs, whereas in area 2 small finger joints are more strongly represented. Area 5 is typically associative in so far as responses in many neurones are obtained only during active movements. The movements activating these neurones are of the extrapyramidal type; these neurones integrate movements and postures of several joints. Thus the movements served by area 5 differ from the precise pyramidal hand and finger movements served by the sensory mechanisms of areas 3, 1 and 2.

The postcentral overall arrangement may thus be conceived as progressing in the anterior to posterior direction from the representation of small elementary sensory events to physically larger patterns that are served by more general and complex mechanisms. Further posteriorly in associative area 7 these mechanisms combine with visual mechanisms for further enlargement of the sphere of sensory control into the immediate extrapersonal space.

B. Parietal Association Cortex

The posterior parietal cortex of the primate has sensory, motor, and behavioural functions that I have recently discussed in a review article from which the following discussion is extended (Hyvärinen, 1982). Useful reviews have also been written by Lynch (1980 b) and Sakata and Kawano (1982).

1. Sensory Functions

a) Visual Functions

It is obvious that the posterior parietal cortex plays a role in visual functions. Evidence for a visual sensory function is derived from studies of posterior parietal lesions in humans and monkeys, and from microelectrode recordings from area 7 in monkeys. In humans posterior parietal lesions produce a visual disorientation syndrome which includes defective orientation in familiar and unfamiliar surroundings, misreaching for targets in the contralateral visual hemifield, constructional apraxia, neglect of contralateral visual hemifield and disturbance of the visual spatial scheme. Defects in monkeys with posterior parietal lesions are less conspicuous, affecting visual orientation in such functions as cage-finding and localization in reference to another object. Monkeys also show the misreaching symptom which is more pronounced in the contralateral visual hemifield during cooling of area 7. In monkeys, too, contralateral visual neglect or extinction occur after cooling or ablation of area 7. As suggested by Semmes et al. (1963), the difficulties in visual orientation after posterior parietal damage could be related to lack of attention to the visual background that consists of the visual patterns surrounding the target of visual inspection. Lack of attention to the background is conspicuous also in "simultanagnosia" in which the patient does not, without provocation, divert his attention from the object he is attending to. The maintained fixity of gaze (Fig. 29A) in Balint's syndrome also reflects inability to change an object in the visual background into the focus of fixation.

Microelectrode recordings show clear visual sensory responses in area 7 (Yin and Mountcastle 1977, Robinson et al. 1978, Motter and Mountcastle 1981). The response latencies are long, the receptive fields of the neurones large, and they often exclude the fovea. Moving stimuli are most often effective, and many neurones have a preferred direction of movement. The preferred directions of movement in the two halves of the visual field may be opposite to each other, resulting in optimal responses to approaching or receding stimuli (Motter and Mountcastle 1981). Variations in the pattern of the stimuli have little effect on the responses. However, the responses are modified by interested visual fixation that augments them (Mountcastle 1981, Mountcastle et al. 1981), and extraretinal signals about eye position (Sakata et al. 1978, 1980, Shibutani et al. 1982).

As suggested by Motter and Mountcastle (1981) the visual sensory properties of area 7 neurones make them suited for a function in the so-called *"ambient visual system"*, as opposed to the *"foveal visual system"* or vision of "object identity" (Trevarthen 1968, Schneider 1969). The ambient visual system operates in the peripheral visual field during foveal work. It can signal visual stimuli moving in the peripheral field enabling interruption of the foveal work when an interesting stimulus enters the peripheral visual field. We can adjust our locomotion and posture on the basis of ambient vision and orient our head and eyes towards targets perceived in this visual system. During locomotion this system keeps us informed of the visual surroundings that are not fixated. The reaching movements of the arms are also performed in peripheral vision while the target is fixated. The ambient vision does not include the space far from the subject and is most efficient in the immediately surrounding space. Area 7 neurones have their receptive fields predominantly in the periphery of the visual field, respond to moving stimuli and enhance their responses during interested visual fixation (Motter and Mountcastle 1981, Mountcastle et al. 1982). These features make these neurones well suited for participation in the ambient visual system. This kind of visual mechanism also fits the role of the cortical component in the cortico-limbic-reticular activating system discussed in Chapter V (pp. 61–62, 68) that shifts attention to visual stimuli moving in the periphery. During our own movements we perceive the surroundings as stable, and are still able to detect peripheral movement that differs from the apparent movement caused by locomotion. The "opponent vector orientation" described by Motter and Mountcastle (1981, see p. 91) suggests that the visual neurones in area 7 may participate in the visual detection of peripheral movement during locomotion.

The counterpart of the ambient visual system is the focal or foveal visual system represented in the occipital cortex and connected to the motor system through the occipito-frontal connections. Interruption of these connections results in poor performance in manual tasks performed in the foveal vision (Haaxma and Kuypers 1974).

The two visual systems, focal and ambient, are a visual analogy of the two motor systems, the "pyramidal" and "extrapyramidal" ones. Indefinite as these systems may be, they emphasize the different roles of the precisely controlled pyramidal movements, such as fine finger movements performed in visual guidance, and the more general postural adjustments executed by the extrapyramidal connections. The posterior parietal lobe is part of the extrapyramidal motor cortex. Thus it is no surprise that the *ambient visual information is used in area 7 for the extrapyramidal motor control whereas the focal visual system can be used to direct many of the fine movements of the pyramidal system.*

b) Somaesthetic Functions

The somatosensory functions of the posterior parietal cortex are just as clear as the visual ones and are likewise reflected in the symptoms caused by lesions in humans and monkeys and in the neuronal recordings from monkeys. In humans the somaesthetic deficits produced by posterior parietal lesions include hemidepersonalization or lack of awareness of the contralateral body half, illusory perception of the opposite body half and asymboly for pain. Contralateral neglect is also reflected in extinction of the contralateral one of two rivalling somaesthetic stimuli. The finger agnosia and right-left disorientation of the Gerstmann syndrome also affect somaesthetic functions.

In monkeys posterior parietal lesions have produced a large number of somaesthetic deficits (Chapter V.B.5, p. 69). Extinction of the contralateral cutaneous stimulus belongs to these as well as various tactile deficits which, however, are less conspicuous than the ones produced by ablation of the primary somatosensory cortex and could be interpreted as an attentional defect related to motor retardation in the contralateral limbs (Ettlinger et al. 1966).

Neuronal responses in the posterior parietal cortex indicate that area 5 is strongly activated by passive joint stimulation and also from the skin. Area 7b is activated by cutaneous and also by kinaesthetic stimuli, and neurones responding to noxious stimuli were described here. In the posterior part of area 7a a kinaesthetic projection area was found. The joint and skin inputs to these neurones are strongly enhanced during voluntary exploration, indicating that the somaesthetic input to posterior parietal neurones may be gated by behavioural variables.

The somatosensory properties of posterior parietal neurones differ
from those of the neurones in the primary somatosensory area. For joint
activation it is typical that passive movements in several joints affect the
activity of area 5 neurones. The cutaneous receptive fields are large and
variable and often directionally selective just as the visual receptive fields
in area 7. It is conceivable that both the joint input and the cutaneous
input to posterior parietal neurones contribute to the sensory control
of posture and movements. The directional cutaneous input could transmit
information about the direction of movement of body parts in relation to
surrounding objects. The input from multiple joints could likewise serve the
motor control of limb and body positions during movements. In that sense
this information could constitute *a somatosensory coordinate system for
goal-directed voluntary movements.*

c) Vestibular and Auditory Functions

The vestibular projection to area 2v is related to somatosensory functions in the head region and serves orientation of the head during movements. The vestibular input to visual tracking neurones and the neurones
responding to optokinetic stimulation in area 7 (Kawano and Sasaki 1981)
signify a vestibular-visual convergence to this region. The vestibular functions can aid orientation, especially in specific situations, e.g., when travel-
ling in a vehicle and in the dark. They may also contribute to eye movement control during head movements. The disturbances in optokinetic
nystagmus after posterior parietal lesions testify to the role of the posterior parietal cortex in functions related to the vestibular system in man,
too.

The lateral and posterior parts of the inferior parietal lobule contain
auditory related cortex in man (area Tpt).Lesions around the posterior
end of the Sylvian fissure produce contralateral auditory inattention in both
species, and neurones in area Tpt of the monkey respond briskly to auditory stimuli. Typical of these neurones is some specificity of the location
of the sound source and sensitivity to natural sound. Neurones related to
head movements also exist in area Tpt and may assist head orientation
towards interesting sounds. In humans speech sounds probably also activate
neurones in this region.

2. Motor Functions

a) Eye Movements

Several types of eye movement disturbances have been described in man after posterior parietal damage. These include fixity of gaze, slowness in voluntary direction of gaze to targets and defective searching eye movements. These symptoms may be viewed as oculomotor apraxia (Cogan 1965). In monkeys lesions of posterior parietal cortex have produced slowing of eye movements towards the contralateral side, slowing of searching eye movements and impairment of tracking eye movements, but the ability to maintain visual fixation has not been disturbed.

The fixity of gaze after posterior parietal damage in man reminds one of the somatic dystonia that follows extrapyramidal cortical ablation. According to Denny-Brown (1966) this dystonia, produced for instance by removal of area 7, represents release reactions of the portion of the motor mechanism that has direct access to alpha motor neurones. Similarly the fixity of gaze ("fixation spasm") after posterior parietal lesion could be viewed as a release of the fixation mechanism from the inhibitory control exercised by posterior parietal cortex. According to this view the role of the posterior parietal cortex is to interrupt fixation when an interesting stimulus appears in the visual periphery.

Electrophysiological evidence also suggests a role of area 7 in eye movement control. Electrical stimulation of area 7a may provoke conjugate contraversive eye movements and stimulation of the posterior part of area 7 pupillary constriction, convergence movements and accommodation. Microelectrode recordings in area 7 have revealed neurones related to visual fixation, tracking and saccades, the fixation neurones being the most numerous of these. As Sakata et al. (1980) showed, the fixation neurones also receive an extraretinal input that signifies eye position in the orbit. Obviously these neurones mediate the action of the posterior parietal cortex on eye movements. The fixation neurones in area 7 probably do not maintain visual fixation but receive a reafference copy of the fixation signal from the frontal eye field or elsewhere, since "fixation neurones" also exist in the prefrontal cortex (Suzuki and Azuma 1977). Such an arrangement would free the posterior parietal cortex for monitoring the peripheral visual field and interrupting fixation upon appearance of interesting peripheral stimuli.

b) Somatic Movements

The disturbances in somatic movements observed in humans after posterior parietal damage are motor neglect of contralateral limbs, misreaching, constructional apraxia, and the agrafia of the Gerstmann syndrome. Monkeys also show paucity of spontaneous movements in limbs contralateral to a lesion, misreach for targets and have difficulty in grasping. Humans with the symptom "optic ataxia" initiate the reaching repeatedly but misreach in the contralateral visual hemifield. After cooling of area 5 monkeys misreached with the contralateral arm in all directions whereas cooling of area 7 resulted in misreaching in the contralateral visual hemifield (Stein 1976, 1978). Thus in the absence of function of the posterior parietal lobe both humans and monkeys are able to initiate reaching movements but the spatial guidance of the movement to the targets is inaccurate.

This result implies that the posterior parietal cortex permutes the information of visual surroundings into a somatic coordinate system for reaching and other movements. These coordinates form a basis for accurate motor programmes for the various muscles participating in reaching and other movements. During the performance of the reach both visual and somaesthetic sensory feedback is used for on-line correction of errors in the movements. Lesions in the posterior parietal cortex destroy the coordinate system for the reach and interrupt the sensory feedback and corollary discharge leading to inaccuracy in performance. The role of areas 5 and 7 differ in this process; area 7 is responsible for the visual part of the coordinate permutation, whereas area 5 further transforms these visual coordinates into a somatic reference system. Microelectrode recordings in area 5 and 7 indicate that neurones related to reaching and manipulation are common in the posterior parietal cortex. The extirpation of this extrapyramidal cortical area leads to dystonia which according to Denny-Brown (1966) represents release of the mechanisms affecting alpha motoneurones. A possible explanation of the role of the reaching and manipulation neurones in the posterior parietal cortex is that they receive the efference copy (von Holst and Mittelstaedt 1950) or corollary discharge (Sperry 1950) from the commands to the muscles producing the movement. Thus extirpation of these neurones does not abolish the movements but leads to defective interplay between the timing of the movement and its spatial guidance.

Constructional apraxia is another form of motor dysfunction observed in humans after posterior parietal damage. This symptom indicates an inability to construct a whole out of parts. It consists of a visual disorientation in details, affects the sphere of hands and fingers and implies an inability to use part of the person (the hand) as a tool in the extrapersonal space (Critchley 1953). This apraxic disorder contains a strong element of agnostic autotopognosia of the hand reminiscent of the finger agnosia of the Gerstmann syndrome with which it often occurs together. The term apractognosia emphasizes the gnostic character of this symptom (Hécaen et al. 1956, Hécaen and Albert 1978). The gnosis or awareness of the praxic actions may be one of the roles of the posterior parietal association cortex in humans rather than the praxic execution itself, which probably depends to a large extent on intact frontal functions. The corollary discharge theory of posterior parietal motor function could easily accommodate lack of awareness of praxis by assuming that an efference copy from the motor commands to hand muscles is sent to the posterior parietal cortex. That the frontal lobe is important in praxic execution was demonstrated in the monkey by Moll and Kuypers (1977), who ablated the premotor cortex in callosotomized monkeys and observed a visuomotor apraxia in the contralateral hand.

c) The Command Hypothesis

Mountcastle et al. (1975) suggested that the functional role of the neuronal machinery in the posterior parietal cortex was to act as a command system for movements in the immediate extrapersonal space. This concept has been strongly criticized ever since it was proposed. The related concept of "command neurones" in invertebrates was critically treated by Kupfermann and Weiss (1978) who considered the concept too vague to be useful in experimental studies (see Fig. 43). The criticism of Goldberg and Robinson (1977) and Robinson et al. (1978) has centered on the fact that the discharge in the visual neurones in area 7 is not predictive of movement but depends on the significance of the stimulus. Accordingly, the discharge may be associated with movement but it may also be dissociated from it. The command concept suits better the frontal motor system; according to new studies of Goldberg and associates, an increase in the discharge of neurones in the frontal eye field is predictive of movement (Bushnell and Goldberg 1979, Goldberg 1980, Goldberg and Robinson 1980, Goldberg and Bushnell 1981).

Humans with an apraxic disorder due to posterior parietal lesion perform certain types of voluntary movements competently in some conditions but are unable to carry them through in other conditions. Thus the command mechanism has remained functional in them, and other factors determine whether they succeed in the task or not. The motor sequence can be commanded when the parietal neurones have been lost (Deuel 1980). Thus all evidence indicates that the functional role of the posterior parietal neurones cannot be based on a simple command concept; even its initial proponents do not seem to emphasize this concept any more (Lynch 1980b, Mountcastle et al. 1980). However, it may not be completely inappropriate to consider the posterior parietal cortex as part of a command system where the sensory and intentional factors converge before decision making.

d) The Corollary Discharge Hypothesis

As von Holst and Mittelstaedt (1950) showed, the principle of efference copy, i.e., a copy of the motor command sent to the sensory structures, could be effectively used for *stabilization of perceptual environment during*

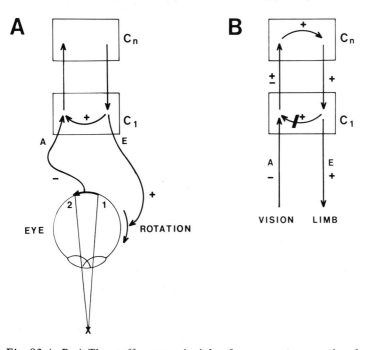

Fig. 83 A, B. A The reafference principle of movement perception during voluntary eye movements. An efferent eye movement command (E) is sent from Center 1 (C_1) for a rotation of the eye to the right (seen from above). A copy marked with a + sign of this efference (reafference) is sent to a sensory structure that also receives an afferent sensory signal indicating that the movement of the eye has caused an apparent leftward movement of a point x in the visual field from 1 to 2 on the retina. This afference (A) has a – sign and is cancelled by the efference copy. Thus other centers (C_n) get no message of movement. This message will only be sent further when there is a mismatch between the two signals. (Modified from von Holst and Mittelstaedt 1950). **B** An extension of the principle of corollary discharge to comprise the reaching movement performed in visual guidance. A motor command for an extension of the arm is sent from a higher level (C_n) to a lower level (C_1) that sends an efferent discharge to the afferent side that also receives a negative signal of the visual perception of the limb movement. When both signals equal no corrective movements are directed from the higher level. However, if the corollary discharge from C_1 to the afferent side is impeded by a lesion (e.g., in the posterior parietal lobe), marked by a *black line*, the afferent signal to C_n is reduced and the feedback from the sensory to the motor side at C_n is reduced leading to a reduction in the efferent command to C_1. Thus the extension of the reaching movement is insufficient.

self-initiated movements. This is illustrated for eye movements in Fig.
83A modified from their paper. When a command is sent from the oculomotor structures (e.g., frontal eye field) to move the eye, for instance, to
the right, an efference copy is simultaneously sent to sensory structures
(e.g., area 7). This efference copy, called corollary discharge by Sperry
(1950), was marked with a + sign and it was neutralized by a sensory
signal with a – sign when the movement caused the sensory signal to be in
the direction opposite to the movement, i.e., in this example to the left.
A movement in the unexpected direction, in this example to the right
would cause a positive signal. At some level in the central nervous system,
which could be, e.g., area 7, the two signals are summed, a positive and
exactly matching negative signal cancelling each other. In that case no
signal of movement arises and the visual surround is perceived stable. The
+ and – signs could be taken for excitation and inhibition and could equally well be reversed. The main requirement for the function of this model
is that the direction of movement is effectively signalled; as we know,
neurones selective for direction of movement are abundant in many parts
of the CNS including the visual and somaesthetic areas and the posterior
parietal lobe. However, if the direction of movement is opposite to the
expected one (caused by some external movement) the two + signals are
additive and a movement sensation arises. An external movement would,
of course, give rise to sensation when no motor command was given.

Placing some posterior parietal neurones at the point of convergence
of the afferent sensory signal and the efference copy could explain several
findings, for instance the inability to control voluntary eye movements
after posterior parietal damage, eye movements caused by strong electrical
stimulation of area 7, e.g., through antidromic conduction to the motor
outflow, sensitivity of area 7 neurones to direction of visual movement,
etc. A perceptual phenomenon presumably caused by the interaction of
an efference copy signal and afferent visual signals is the illusory sigma-
movement produced by simultaneoulsy flickering stationary periodic
stimuli. The perception of this phenomenon is disturbed in posterior
parietal and cerebellar lesions (Buettner et al. 1982), suggesting that these
structures play a role in the corollary discharge mechanism. However, as
shown by von Holst and Mittelstaedt (1950), the principle of reafference
or efference copy is probably in wide use in fairly elementary nervous
systems such as isolated ganglia of invertebrates. Moreover, it can be
modified for the explanation of involuntary movements such as e.g.,
optokinetic nystagmus. Thus the parietal cortex is probably not the
only site in the CNS where it is operative.

In the somaesthetic sense an example of the corollary discharge
is given by the muscle spindle with its sensory and gamma-motor inner-

vations. The motoneuronal alpha-gamma linkage provides automatically
an efference copy via the spindle afferents whose activity is modulated
if peripheral resistance impedes the intended movement (Granit 1972).
However, in humans with peripheral muscular paralysis willed efforts to
move the paralyzed body part do not lead to perception of movement
which is taken as evidence for lack of corollary discharge (McCloskey and
Torda 1975). However, peripheral paralysis with curare-type drugs does
not dissociate the alpha-gamma linkage and does not disprove the possible involvement of a peripheral corollary discharge mechanism which
could make use of the gamma-efferent system and thus alleviate the need
for a central efference copy. The dissociation of the alpha-gamma linkage
by vibration of muscle tendons can produce illusions of movement (Goodwin et al. 1972) indicating that the increased spindle afferent inflow
could serve as a positive corollary discharge which in the absence of nega-
tively coupled joint input could cause an illusion of movement. (For
further discussion see Matthews 1982).

The misreaching symptom produced by parietal lesions could also be
explained by assuming that the path of a corollary discharge has been
interrupted by the lesion. For the performance of the reach a command is
sent to the extensors of the arm, and the extent of the movement is
limited by visual signals indicating successful performance. However,
as indicated in Fig. 83B, the absence of efference copy would cause the
extension to be insufficient in comparison with the visual information
indicating reasonable performance. This mismatch would lead to a decrease in the commanded effort and the extension movement. The reduced amplitude of the extension would then cause a delayed attempt for
correction leading perhaps to overextension followed by a compensatory
flexion which again could be similarly affected by the lack of the efference copy or corollary discharge.

This reasoning suggests that the role of the posterior parietal lobe in
motor performance is related to the mechanism of corollary discharge.
However, as stated above, the posterior parietal cortex is certainly not the
only structure related to the corollary discharge but only one of many
sites where this principle may operate. The posterior parietal lobe could,
anyway, represent a critical locus in the circuit of corollary discharge for
certain types of movements such as the reaching movements performed in
visual guidance, and searching eye movements. This basic concept leaves
many problems open; one of these is, for instance, the mechanism of
visual estimation of the accuracy of the limb movement, but such an
analysis could preferably be considered a function of the visual areas.

Participation in the circuit of corollary discharge is probably only
one of the functions of the posterior parietal lobe. However, this function

seems to suit the role of this cortical area in motor performance better than does command function.

3. Behavioural Functions

a) Sensorimotor Interaction

Traditionally, physiological research of the cerebral cortex has been divided into two parts: the study of sensory systems and the study of motor systems. This dichotomy has also been applied to studies of posterior parietal cortex. Mountcastle et al. (1975) and Lynch et al. (1977) considered that neurones in the posterior parietal cortex act as a "command system" for movements in extrapersonal space, whereas Robinson et al. (1978) regarded all functions of area 7 as sensory. This argument has recently been reviewed by Lynch (1980b).

It is of conceptional interest to determine whether a neurone is sensory or motor. However, it is questionable whether such a dichotomous approach is valid for the associative cortex in non-anaesthetized, behaving

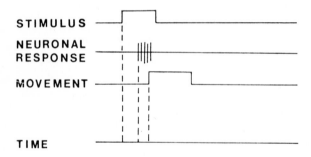

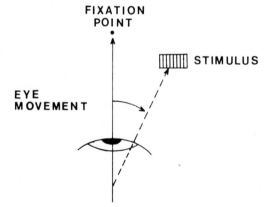

Fig. 84. The relationship of sensory stimulus, neuronal response and behavioural response in a simple reflex arc that leads from visual stimulus to eye movement. In the parietal association cortex the functional coupling from the stimulus to the response becomes conditional as well as the coupling between the discharge and the behavioural response. (Hyvärinen 1980)

animals or man. We may consider an eye movement as an example. An eye movement does not occur without a reason; it is triggered by some external stimulus or by internal expectations. A reflex mechanism that turns the gaze towards a visual stimulus is conceivable. For that purpose the discharge evoked by the stimulus must be coupled to those eye muscles that turn the gaze in the direction of the stimulus. In such a reflex circuit the sensory stimulus naturally precedes the discharge of the neurones and this discharge precedes the eye movement (Fig. 84, Hyvärinen 1980, 1981b).

However, on the level of cortical associative information processing such simple reflexes are not likely to occur. Here eye movements are matched with the behavioural context; the occurrence of the eye movement depends on an analysis of the character of the stimulus and an evaluation of its importance in motivational terms. If such an analysis suggests that the stimulus differs significantly from the background, mechanisms of attention are triggered into operation in the peripheral visual field. If the stimulus is judged important, an eye movement is made towards the stimulus which by now has been differentiated from the background and made a target of search. Much of this analysis is made outside foveal vision, whereas foveal vision is used for closer inspection of the target.

In such a system neurones or neuronal ensembles are not merely sensory or merely motor; their essential role is the integration of these aspects. Therefore, it is not surprising that area 7 consists of neurones with different functional properties; some are more directly related to sensory input, some to motor output, and some combine both these mechanisms. As a whole these neurones comprise part of a machinery that is able to interact between outside stimuli and motor control. Motor control is based on internal purpose and sensory information (Granit 1977). Information of purpose and sensory targets has to reach the elements of this machinery either directly or through various intermediary stages (that also include memory traces) which precede the synaptic input to these neurones. Since a voluntary movement is directed in space there must be an input system to these neurones that is capable of transmitting information about direction in space. That information can only be extracted from sensory cues.

The parietal cortical machinery appears to perform much of the analysis of direction of movement in the surrounding near space although it makes use of multiple contacts with other areas of the brain for this purpose. Therefore, this area contains neurones related to various aspects of this task, and most probably neurones with similar properties also exist in other areas, for instance in the premotor cortex (Hyvärinen 1980, 1981b).

It now appears that the opinions of various researchers of posterior parietal cortex have come closer to each other with understanding of the interactive nature of the information processing in the association cortex. This is evident in the discussion following the review of Lynch (1980b), as well as in private discussions.

Examples of sensorimotor interaction in the posterior parietal cortex are seen in neuronal responses during active movements. Joint neurones in area 5 discharge more vigorously during active than passive movements (Mountcastle et al. 1975), and arm projection and hand manipulation neurones probably receive sensory signals from joints, muscles and skin during active movements. It appears likely that the sensory activity in these neurones is "gated" by inputs related to the preparation of the motor act. A possible mechanism for such action could be a "gate opening" conditioning input in the form of corollary discharge from motor structures. Similar mechanisms may operate in the sphere of vision during active looking (Hyvärinen and Poranen 1974, Sakata et al. 1980, Mountcastle et al. 1982) and in active touch (Hyvärinen and Poranen 1978b, Hyvärinen 1981c).

It is also likely that some neurones are more closely related to the sensory part of the processing while others are involved in the motor aspect. As emphasized by Motter and Mountcastle (1981), in the posterior parietal cortex there probably exists a spectrum of neurones ranging from those activated by passive stimuli and unaffected by movements, through those with combined properties, to those active during movements and insensitive to passive stimulation. In such a system intention, attention and motivation may interact with the sensorimotor processing, enabling the processing to pass all the way to action only when the stimulus is related to the intentions of the animal or man.

b) Spatial Schema

When we are mobile we have no stable cues for the spatial structure of our environment; yet we perceive the environment as a stable system. It is obvious that for stable spatial awareness all the different kinds of cues must be utilized simultaneously. By the combination of different kinds of visual, somatic, vestibular, and auditory cues a reasonably stable spatial reference system can be constructed. A powerful tool for the perceptual stabilization of the surrounding space is the mechanism of corollary discharge discussed above. Various conceivable reafference loops for the stabilization of signals within and across different modalities might occupy a significant share of posterior parietal cortical tissue.

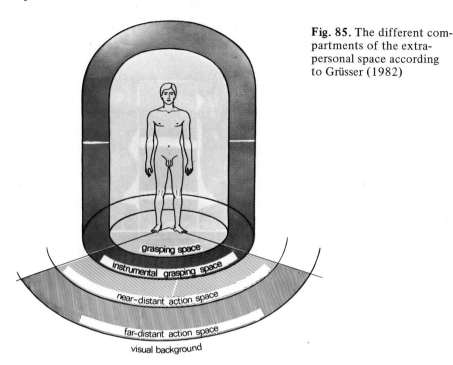

Fig. 85. The different compartments of the extrapersonal space according to Grüsser (1982)

As shown by Grüsser (1982), the space around an individual can be considered to be composed of functionally different sub-spaces (Fig. 85). Outside the actual body space, but closest to the individual, there is the grasping space which is surrounded by an instrumental extension of this space. Further away follow the near-distant action space, far-distant action space, and visual background. Different strategies apply for operations in these different parts of the surrounding space. It may well be that some specialization of the neuronal machinery may have developed for operations in each of these parts of the surrounding space. In some patients the spatial disorder may be limited e.g. only to the near grasping space, whereas in others operations in the more distant parts of the surrounding space may be more involved.

Normally the most important spatial cues derive from the structure of the visual world, but blind people and normal people in the dark can synthesize a representation of space from somatic vestibular and auditory cues. One of the tasks of the posterior parietal cortex is apparently to maintain a spatial reference system which is used for guiding movements in extrapersonal space (Hyvärinen and Poranen 1974). The properties of the visual sensory neurones make them well suited for this task. The visual spatial reference system is based on an analysis of movements in the visual background while the foveal vision is focused on a target. Area 7 visual neurones are sensitive to movements in the peripheral visual field, and interested fixation further enhances their sensitivity (Motter and Mountcastle 1981, Mountcastle et al. 1982). Neurones in area 7b combine visual and somatic information in a way that suggests their use as a spatial reference system which is to some extent free from the constraints of one modality (see Chap. VIII.C). For instance, the visual activating region may be close to the somatic receptive field, and the effective direction of stimulus movement may be identical in both modalities (Hyvärinen and Poranen 1974, Leinonen et al. 1979, Leinonen and Nyman 1979, Leinonen 1980). The auditory neurones in area Tpt likewise process crude specificity to location of the sound source, and sounds are often effective stimuli when they derive from the same direction where the somatic receptive fields are located (Leinonen et al. 1980, see Chap. XI.B). This kind of convergence of information from several sources in multiple elements could form a stable spatial reference system which is continually updated during movements. The multiplicity of anatomical connections (see Chap. IV) also lends support to the use of multiple cues in the function of the posterior parietal cortex.

Lesion data from both humans and monkeys support the notion of a representation of extrapersonal space in the posterior parietal cortex. In humans visual disorientation, simultanagnosia, inadequate understanding of maps, misreaching, and constructional apraxia all suggest a defect in the understanding of the structure of visual space. A defect in the representational spatial schema was also suggested in the studies of Bisiach et al. (Bisiach and Luzzatti 1978, Bisiach et al. 1979), who found that posterior parietal damage caused neglect of the contralateral half of the mental representation of space. Furthermore, the defect in somatic awareness of contralateral body half, autotopagnosia, finger agnosia, and right-left disorder suggest that the spatial defect affects also the understanding of one's own body position in relation to the surroundings.

The lesion studies in monkeys do not indicate equally clear spatial defects as those found in man (see Chap. V.B). However, there is evidence of visuo-spatial disorientation, defective allocentric location, misreaching, and contralateral somatic defects in monkeys, suggesting that the posterior parietal cortex may have a function related to the understanding of the contralateral space in non-human primates, too.

On the basis of the first microelectrode recordings of neuronal activity in the posterior parietal cortex I concluded in 1974 that "sensory signals relayed with reference to spatial locations to the cells of the posterior association cortex can be used to guide the motor systems during voluntary movements aimed at targets placed at various spatial locations"

(Hyvärinen and Poranen 1974). The view of a central role of the posterior parietal lobe in spatial perception now appears generally accepted (Sakata et al. 1980, Leinonen 1981, Motter and Mountcastle 1981).

c) Motivation-Intention-Attention

The posterior parietal cortex is connected inter alia with limbic and hypothalamic structures (see Chap. IV). Thus it is conceivable that *motivational factors* could influence the activity. However, lesion studies in humans and monkeys do not reveal appreciable disturbances in motivation, and the recordings of Rolls et al. (1979, 1980) likewise indicate that motivation as such is not a strong factor in activating visual fixation neurones in area 7 of the monkey. The motivational influences from limbic and hypothalamic structures may, however, participate as an additional factor in the responses related to intentional and attentional processes.

The influence of the *intention* of the animal on the firing of neurones in the posterior parietal cortex is clearly more substantial than the influence of motivation. A strong intentional influence is evident in the activity of the arm projection and hand manipulation neurones and the facilitation of activity in the visual sensory neurones by interested fixation. A defect in intention rather than attention was also postulated as the basis of the contralateral neglect syndrome in monkeys by Watson et al. (1978), who found that monkeys with contralateral neglect perceived the contralateral stimulus but did not respond to it (see Chap. V.B.4). However, the lesions in these animals were frontal or reticular; without investigating monkeys with parietal lesions it cannot be concluded for sure that parietal lesions in monkeys also result in an intentional deficit. An intentional deficit in man was suggested by Heilman and Valenstein (1979), who considered that a unilateral hypokinesia best explained the contralateral neglect ensuing in human patients from right parietal and other lesions.

The relationship between intention and attention is close. Attention towards a stimulus follows intention, orientation and categorized search for the proper stimulus and is often succeeded by action. Thus the behaviours in the chain *intention-orientation-attention-action* are closely related, and they all affect posterior parietal function. The interpretation of the role of associative cortical neurones in this behavioural chain is bound to be arbitrary.

The role of *attention* in posterior parietal function is documented in many studies. Significance of posterior parietal cortex for attention is indicated by the syndrome of contralateral neglect in humans and the contralateral extinction in monkeys. The defect in humans called simultanagnosia or inability to observe several phenomena simultaneously also

indicates a defect in the mechanism that shifts attention from one target
to another.

Neuronal recordings also suggest that attention plays a role in function
of posterior parietal cortex. Bushnell et al. (1978, 1981) demonstrated
that the activity of area 7 neurones is enhanced prior to a saccade to the
receptive field of the neurone, and this enhancement also occurs when the
animal attends to the stimulus without making a saccade. Similar enhancement occurred in the frontal eye field and the superior colliculus
only in connection with saccades to the receptive field (Bushnell and
Goldberg 1979, Goldberg 1980, Goldberg and Robinson 1980, Robinson
and Keys 1980, Goldberg and Bushnell 1981), whereas in the pulvinar
enhancement of activity preceded any eye movements and not only those
to the receptive field (Keys and Robinson 1979, Robinson and Keys
1980). These results led Goldberg and Robinson (1980) to conclude that,
contrary to other parts of the visuo-motor system, neurones in area 7 give
an index of spatial location, and when their activity is enhanced, indicate
that the object in the receptive field is important for the animal. On the
other hand, frontal eye field neurones, which are related to specific spatial
locations, when enhanced always signify an impending eye movement
only.

The results of Mountcastle et al. (1981) also indicate that attention
toward visual signals is important for the activity of the visual sensory
neurones in area 7. They showed that interested visual fixation greatly
enhances responses to passive visual stimuli on the receptive field, but
they did not find enhancement prior to saccades. Such enhancement
during interested visual fixation could be a sign of selective visual attention
towards visual stimuli.

Watson and Heilman et al. (Watson et al. 1973, 1974, Heilman and
Watson 1977a,b, Heilman et al. 1978) suggested that neglect of contralateral stimuli follows interruption of a cortico-limbic-reticular activating
system. According to this theory significant stimuli could lead to activation of the ascending reticular system through a connection via the limbic
cortex in the cingulum. Such an arrangement could be part of the normal
mechanism of focussing attention; the role of the parietal cortex in attention is probably in its spatial direction.

d) Plasticity, Learning, Memory

The data presented in Chap. XI show that the posterior parietal mechanisms are modifiable at young age. When monkeys are prevented from
using vision during the first months of their lives its representation declines
in association cortex; the deficit appears to be lasting. At the same time
mechanisms related to use of somatic modalities gain in representation

(Hyvärinen et al. 1981a,b,c). Such findings show that abnormal procedures may alter the organization of the associative cortex.

Less dramatic changes are likely to occur during normal development and learning. For instance in infancy we learn to use our hands in visual guidance (McDonnell 1979). The perfection of this function requires a lot of training at the appropriate age. Lack of this training blocks the normal development of visuo-motor coordination, as is shown by infant monkeys prevented from seeing their hands (Held and Bauer 1967). At toddler age we learn to move in the surrounding space. During the period of locomotion training the mechanisms that stabilize the perceptual environment mature. The characteristic unsteady locomotion of a toddler may depend in part on the immaturity of the neural system that stabilizes the perceptual environment during self-initiated movements by use of corollary discharge or other mechanisms. The systems involved in perceptual feedback and stabilization require an extensive training period which could be compared with the period of learning language. During this period of learning the neural mechanisms in the parietal lobe and elsewhere gradually consolidate to form the mature functional system.

In the mature nervous system learning is likely to affect posterior parietal mechanisms in more subtle ways. Memory engrams and learning affecting the posterior parietal lobe occur in such familiar tasks as reaching towards moving targets, orientation in changing surroundings, drawing, construction, etc., or for that matter the experimentally conditioned eye or hand movement towards a peripheral target. For instance, we can memorize even in the dark the spatial localization of targets learned through use of vision and this can be done after a considerable time lapse from the learning occasion. Moreover, blind people can construct their mental representation of surrounding space entirely through serial somatic and auditory observations retained in memory. In fact the performance of simple daily activities requires that the underlying neural systems have a memory function embedded in them. The requirement of a reliable memory mechanism thus concerns the posterior parietal association cortex as much any other part of the brain.

4. Cellular Machinery

The reason for considering the functional role of posterior parietal lobe is that it might provide understanding of the physiological mechanism by which this piece of tissue achieves that role. If we understand the functional role clearly we can ask the right questions about the mechanism through which the purpose is fulfilled. The main features known of this mechanism are dealt with in this book. However, against the background of its functional role our knowledge of the cellular mechanism of the posterior parietal cortex appears still quite fragmentary. We know now the main features of the functional organization and some of the input mechanisms affecting cellular activity in this region. Many of the inputs required for fulfillment of the role of the posterior parietal cortex have not yet been seen and many have not even been looked for. Moreover, conceptualization of the experiments that could reveal some of them indicates formidable methodological difficulties.

a) Functional Organization

The organization of posterior parietal cortex appears to be based on the three main systems present everywhere in the cortex: the *regional, laminar, and columnar organizations.* Within the posterior parietal cortex there are subregions that have clearly different functions. Area 5 is almost totally devoted to somatic sensory mechanisms of posture and movement. The anterior part of area 7a is mainly related to vision and eye movements, whereas the posterior part of it also has kinaesthetic functions. The most lateral part of area 7b is mainly somaesthetic, whereas the more medial parts of it contain both somatic and visual mechanisms. Area Tpt is mainly auditory, and area 2v (and possibly POa) contains vestibular mechanisms. Therefore, it is important to localize accurately the site of study within the posterior parietal cortex of the monkey. Presumably such areal differences also exist in the human posterior parietal cortex and contribute to the variability of symptoms following lesions.

Little is known so far about columnar and laminar organization of posterior parietal cortex. It is possible that the different kinds of neurones observed from intracortical processing chains leading from sensory input to motor output (Motter and Mountcastle 1981). Vertical columnar processing chains with functionally different layers probably exist but will have to wait for proof. In view of the multiplicity of the anatomical connections of the posterior parietal cortex the processing in the columnar and laminar structures is likely to be quite complex. The purpose of this complex system of connections may be to synthesize a spatial representation from different inputs, each of which alone carries fairly little information of space, and to combine with this representation factors signifying intention, attention and movement in the form of corollary discharge. Thus the cellular machinery in the posterior parietal cortex *combines information of personal (somatic) awareness with extrapersonal (largely visual) space and internal purpose with external stimuli.* Such a machinery must deal with behavioural mechanisms, two of which are personal in nature, personal awareness and personal intention, and two externally oriented, the representation of external space and attention towards external stimuli. The main task of the posterior parietal cortex appears to be the proper combination of this information for the use in guidance of behaviour in rapidly changing situations.

b) Methodological Difficulties

The classification of neurones studied by different authors in the posterior parietal cortex has varied greatly. Much of this variation depends on functional differences between different parts of the posterior parietal cortex, but another source of variability is the type of classification adopted by different authors. When the neurones under study have variable functional properties, their classification may be misleading if it is based on only one property, such as relationship to eye movements or to visual sensory stimuli. When such classification is used many neurones cannot be classified in any category; their function is not directly related to the variables in the classification. For these reasons the prevalence of different neurone types in different studies may vary largely depending on the design of the experiment. As Mountcastle has shown in three successive studies from his laboratory, variation of objectives and methods leads to great variability of the percentages of neurones related to fixation, eye movements, visual stimuli, arm projection and manipulation (Motter and Mountcastle 1981). In other laboratories quite different classifications could be adopted, as shown by Robinson et al. (1978). The classification of neurones with such complex and variable properties as those in area 7 is always likely to be erroneous in some functional aspects. Therefore, qualitative studies in which the nature of neuronal function can be investigated with a large number of flexibly variable tests still constitute the most reliable method of disclosing the over-all percentages of different neurone types in various parts of the posterior parietal cortex. Studies enabling the repetition of the same observation several times under quantitative control are, of course, useful in revealing closer relationships of neuronal action to sensory stimuli and motor behaviour. For the advancement of knowledge of association cortex both types of studies are needed.

c) Factors That Influence Cellular Activity

According to the analysis of the functional role the factors mentioned in Sections B.1–B.3 should have access to information processing in the neurones of the posterior parietal cortex. In view of the multitude of neural connections of this cortical region they are all likely to do so. We have already a considerable amount of information on the sensory factors. Visual sensory mechanisms related mainly to peripheral vision and sensitive to the direction of movement were described by Motter and Mountcastle (1981). Visual fixation and eye position also influence the activity of area

7 neurones (Sakata et al. 1980, Mountcastle et al. 1982). However, we know nothing yet of the mechanism that differentiates the target of fixation from the visual background, a function likely to be largely occipital but affecting the neurones in area 7, too. The parietal mechanism in the motor act of foveation and focussing also remains open although spatial direction of visual attention is implicated.

The somaesthetic functions of area 7b appear related to awareness of personal space and to localization of external stimuli in relation to the head, body and limbs, whereas the somatic mechanisms in area 5 are related to guidance of limb movements towards targets. The mechanisms of sensory control of the movements remain open, however, as well as the existence and circuitry of an efference copy in this area.

The vestibular functions in the parietal lobe probably serve orientation of the head and perceptual stability during head and eye movements. The auditory neurones in area Tpt also appear to signal the structure of the surrounding space. This they do by responding differentially, depending on the location of the sound source. The acoustic determinants of this function remain open, however.

The nature of the relationship of posterior parietal activity to motor mechanisms is unclear. The command function does not appear as a likely role for most parietal neurones, whereas the corollary discharge or efference copy mechanism may be important as suggested by the capacity of this mechanism to stabilize the perceptual environment during locomotion and movements. The corollary discharge mechanism is only a hypothesis; so far there is no experimental basis for it on the level of cellular activity in the parietal lobe. The testing of this hypothesis could most easily be performed with regard to eye movements that can be performed when the head is fixed without causing disturbances in the recordings. On the other hand, cellular recordings performed during self-induced locomotion face formidable difficulties.

The behavioural factors that might influence posterior parietal activity have been examined only quite superficially. Various factors could augment or reduce the responses of posterior parietal neurones and thus act as modulating "gating" inputs. Motivation as such is perhaps not a very important modulator but intention and attention are likely to have a significant influence on cellular activity in the posterior parietal lobe. Moreover, this influence may follow certain rules in terms of laminar and columnar organization. Experiments for revealing the influence of such factors are not impossible to design, but carrying them through with quantitative techniques is a major undertaking.

Cellular recordings related to the spatial representational scheme, learning, and memory are conceptually difficult at this phase of research. However, we may always rely upon the creativity of the mind of the individual experimenter for making startling new discoveries even about these difficult topics.

C. Parietal Lobe as a Whole

The anatomical boundaries of the parietal lobe are distinct anteriorly (central sulcus) and laterally (Sylvian fissure). In monkeys the posterior boundary (superior temporal sulcus) is also clear, whereas in humans the boundary with the occipital lobe does not consist of a uniform sulcal margin. Thus in humans the posterior parietal lobe and the occipital circumstriate belt are not sharply demarcated. The parietal lobe contains functionally different zones such as the primary somatosensory area, area 5, and area 7. At first glance it might seem that there is no common functional feature for the whole parietal lobe. However, a certain degree of coherence can be seen in its various functions. Anteriorly, the somatosensory cortex mediates accurate information concerning cutaneous stimuli and joint positions. This information is used for accurate pyramidal guidance of somatic movements and exploration. Posteriorly this information is used in area 5 for somatic sensory guidance of larger-scale movements directed by the extrapyramidal system. The inferior parietal lobule receives visual input in addition to the somatic one. This visual input is related to positions in the extrapersonal space and, particularly in area 7b, to the somatic representation of the body and its parts. Data serving the broad-scale somatosensory awareness are transmitted to the inferior parietal lobule at least partly from the second somatosensory area in the parietal operculum. All the parts of the parietal lobe are thus linked with a system that serves perception of the body and its parts in relation to external space and the guidance of movements of the body and its parts, including the eyes, to targets of interest in that space. Different parts of the parietal lobe serve different body parts and different functions within this common parietal system. In terms of visual-oculomotor functions, the analysis of the body and its parts and their guidance does not differ distinctly from visual-oculomotor tasks related to external space. Thus a gradual shift between the occipital and the parietal visual functions is natural and is perhaps reflected in the lack of a sharp anatomical boundary between the parietal and occipital lobes of the human brain.

References

Ackroyd C, Humphrey NK, Warrington EK (1974) Lasting effects of early blindness. A case study. Q J Exp Psychol 26:114-124

Adey WR, Segundo JP, Livingstone RB (1957) Corticofugal influence on intrinsic brainstem conduction in cat and monkey. J Neurophysiol 20:1-16

Adrian ED (1941) Afferent discharge to the cerebral cortex from peripheral sense organs. J Physiol 100:159-191

Albe-Fessard D, Fessard A (1963) Thalamic integrations and their consequences in the telencephalic level. Progr Brain Res 1:115-148

Allison RS, Hurwitz LJ, White JG, Wilmot TJ (1969) A follow-up study of a patient with Balint's syndrome. Neuropsychologia 7:319-333

Altrocchi PH, Menkes JH (1960) Congenital ocular motor apraxia. Brain 83:579-588

Awaya S, Miyake Y, Imaizumi Y, Shiose Y, Kanda T, Komuro K (1973) Amblyopia in man, suggestive of stimulus deprivation amblyopia. Jpn J Ophthalmol 17:69-82

Baker FH, Grigg P, Noorden von GK (1974) Effects of visual deprivation and strabismus on the response of neurons in the visual cortex of the monkey, including studies on the striate and prestrate cortex in the normal animal. Brain Res 66: 185-208

Balint R (1909) Seelenlähmung des "Schauens", optische Ataxie, räumliche Störung der Aufmerksamkeit. Monatsschr Psychiatr Neurol 25:51-81

Barlow HB, Levick WR (1965) The mechanism of directionally selective units in rabbit's retina. J Physiol 178:477-504

Bartholow R (1874) Experimental investigations into the functions of the human brain. Am J Med Sci 67:305-313

Bates JAV, Ettlinger G (1960) Posterior biparietal ablations in the monkey. Arch Neurol 3:177-192

Baxter BL (1966) The effect of visual deprivation during postnatal maturation on the electrocroticogram of the cat. Exp Neurol 14:224-237

Bechterew von G (1911) Die Funktionen der Nervencentra. Heft 3. Fischer, Jena

Bender MB, Furlow LT (1944) Phenomena of visual extinction and binocular rivalry mechanisms. Trans Am Neurol Assoc 70:87-92

Bender MB, Teuber HL (1947) Spatial organization of visual perception following injury to the brain. Arch Neurol Psychiatr 58:721-739, 59:39-62

Benevento LA, Rezak M (1976) The cortical projections of the inferior pulvinar and adjacent lateral pulvinar in the rhesus monkey (macaca mulatta): an autoradiographic study. Brain Res 108:1-24

Benson DF, Geschwind N (1970) Developmental Gerstmann syndrome. Neurology 20:293-298

Benton AL (1961) The fiction of the "Gerstmann syndrome". J Neurol Neurosurg Psychiatr 24:176-181

Berger H (1900) Experimentell-anatomische Studien über die durch Mangel optischer Reize veranlassten Entwicklungshemmungen im Occipitallappen des Hundes und der Katze. Arch Psychiatr Nervenkr 33:521-567

Bertrand M (1969) The behavioural repertoire of the stumptail macaque. Karger, Basel, pp 232-254

Bianchi L (1895) The functions of the frontal lobes. Brain 18:497-522

Bisiach E, Luzzatti C (1978) Unilateral neglect of representational space. Cortex 14: 129-133

Bisiach E, Luzzatti C, Perani D (1979) Unilateral neglect, representational schema and consciousness. Brain 102:609-618

Blakemore C, Sluyters van RC (1975) Innate and environmental factors in the development of the kitten's visual cortex. J Physiol 248:663-671

Blakemore C, Garey LJ, Vital-Durand F (1978) The physiological effects of monocular deprivation and their reversal in the monkey's visual cortex. J Physiol 283:223-262

Bonin von G, Bailey P (1947) The neocortex of Macaca mulatta. University of Illinois Press, Urbana

Brain WR (1941) Visual disorientation with special reference to lesions of the right cerebral hemisphere. Brain 64:244-272

Brodmann K (1905) Beiträge zur histologischen Lokalisation der Grosshirnrinde. Dritte Mitteilung: Die Rindenfelder der niederen Affen. J Psychol Neurol 4:177-226

Brodmann K (1907) Beiträge zur histologischen Lokalisation der Grosshirnrinde. Sechste Mitteilung: Die Cortexgliederung des Menschen. J Psychol Neurol 10:231-246

Brugge JF, Merzenich MM (1973) Responses of neurons in auditory cortex of the macaque monkey to monaural and binaural stimulation. J Neurophysiol 36:1138-1158

Buettner UW, Dichgans J, Grüsser O-J (1982) Efferent motion perception (δ-movement) and δ-pursuit in neurological patients. In: Lennerstrand G, Zee D, Keller EL (eds) Functional basis of ocular motility disorders. Pergamon Press, Oxford, in press

Büttner U, Henn V (1976) Thalamic unit activity in the alert monkey during natural vestibular stimulation. Brain Res 103:127-132

Büttner U, Buettner UW (1978) Parietal cortex (2v) neuronal activity in the alert monkey during natural vestibular and optokinetic stimulation. Brain Res 153:392-397

Buisseret P, Imbert M (1976) Visual cortical cells: Their developmental properties in normal and dark reared kittens. J Physiol 255:511-525

Burton H, Jones EG (1976) The posterior thalamic region and its cortical projection in New World and Old World monkeys. J Comp Neurol 168:249-302

Bushnell MC, Goldberg ME (1979) The monkey frontal eye fields have a neuronal signal that precedes visually guided saccades. Neurosci Abstr 5:779

Bushnell MC, Goldberg ME, Robinson DL (1978) Dissociation of movement and attention: neuronal correlations in posterior parietal cortex. Neurosci Abstr 4:621

Bushnell MC, Goldberg ME, Robinson DL (1981) Behavioral enhancement of visual responses in monkey cerebral cortex. I. Modulation in posterior parietal cortex related to selective visual attention. J Neurophysiol 46:755-772

Cajal S Ramón y (1937) Recollections of my life. Craigie EH (transl). Am Philos Soc, Philadelphia, p 534

Carmichael EA, Dix MR, Hallpike CS (1954) Lesions of the cerebral hemispheres and their effects on optokinetic and caloric nystagmus. Brain 77:345-372

Chavis DA, Pandya DN (1976) Further observations on corticofrontal connections in the rhesus monkey. Brain Res 117:369-386

Chow KL, Hutt PJ (1953) The "association cortex" of Macaca mulatta: A review of recent contributions to its anatomy and function. Brain 76:625-677

Cogan DG (1953) A type of congenital motor apraxia presenting jerky head movements. Am J Ophthalmol 36:433-441

Cogan DG (1965) Ophthalmic manifestations of bilateral non-occipital cerebral lesions. Br J Ophthalmol 49:281-297

Cogan DG, Adams RD (1953) A type of paralysis of conjugate gaze (ocular motor apraxia). Arch Ophthalmol 50:434-442

Cogan DG, Adams RD (1955) Balint's syndrome and ocular motor apraxia. Arch Ophthalmol 53:758

Cogan DG, Loeb DR (1949) Optokinetic response and intracranial lesions. Arch Neurol Psychiatr 61:183-187

Cole M, Schutta HS, Warrington EK (1962) Visual disorientation in homonymous half-fields. Neurology 12:257-263

Costanzo RM, Gardner EP (1980) A quantitative analysis of responses of direction-sensitive neurons in somatosensory cortex of awake monkeys. J Neurophysiol 43:1319-1341

Crawford MLJ, Blake R, Cool SJ, Noorden von GK (1975) Physiological consequences of unilateral and bilateral eye closure in macaque monkeys: some further observations. Brain Res 84:150-154

Critchley M (1953) The parietal lobes. Arnold, London

Critchley M (1966) The enigma of Gerstmann's syndrome. Brain 89:183-199

Damasio AR, Benton AL (1979) Impairment of hand movements under visual guidance. Neurology 29:170-178

Deecke L, Schwarz DWF, Fredrickson JM (1977) Vestibular responses in the rhesus monkey ventroposterior thalamus. II. Vestibulo-proprioceptive convergence at thalamic neurons. Exp Brain Res 30:219-232

Denny-Brown D (1966) The cerebral control of movement. Liverpool University Press, Liverpool

Denny-Brown D, Banker B (1954) Amorphosynthesis from left parietal lesion. Arch Neurol Psychiatr 71:302-313

Denny-Brown D, Chambers RA (1958) The parietal lobe and behavior. Res Publ Assoc Res Nerv Ment Dis 36:35-117

Denny-Brown D, Meyer JS, Horenstein S (1952) The significance of perceptual rivalry resulting from parietal lesion. Brain 75:433-471

Deuel RK (1977) Loss of motor habits after cortical lesions. Neuropsychologia 15:205-215

Deuel RK (1980) The parietal association fields and motor behavior. Behav Brain Sci 3:501

Deuel RK, Collins RC, Dunlop N, Caston TV (1979) Recovery from unilateral neglect: Behavioral and functional anatomic correlations in monkey. Neurosci Abstr 5:624

De Renzi E, Faglioni P, Scotti G (1970) Hemispheric contribution to exploration of space through the visual and tactile modality. Cortex 6:191-203

DiPerri R, Dravid A, Schweigerdt A, Himwich HE (1968) Effects of alcohol on evoked potentials of various parts of the central nervous system of cat. Q J Stud Alcohol 29:20-37

Divac I, Lavail JH, Racic P, Winston KR (1977) Heterogeneous afferents to the inferior parietal lobule of the rhesus monkey revealed by the retrograde transport method. Brain Res 123:197-207

Domino EF (1964) Neurobiology of phencyclidine (Sernyl), a drug with an unusual spectrum of pharmacological activity. Int Rev Neurobiol 6:303-347

Drewe EA, Ettlinger G, Milner AD, Passingham RE (1970) A comparative review of the results of neuropsychological research on man and monkey. Cortex 6:129-163

Duffy FH, Burchfield JL (1971) Somatosensory system: Organizational hierarchy from single units in monkey area 5. Science 172:273-275

Economo von C (1929) The cytoarchitectonics of the human cerebral cortex. Oxford University Press, London

Essen van DC (1979) Visual areas of the mammalian cerebral cortex. Annu Rev Neurosci 2:227-263

Ettlinger G (1977) Parietal cortex in visual orientation. In: Rose FC (ed) Physiological aspects of clinical neurology. Blackwell, Oxford, pp 93-100

Ettlinger G, Kalsbeck JE (1962) Changes in tactile discrimination and in visual searching after successive and simultaneous bilateral posterior parietal ablations in the monkey. J Neurol Neurosurg Psychiatr 25:256-268

Ettlinger G, Warrington E, Zangwill OL (1957) A further study of visual-spatial agnosia. Brain 80:335-361

Ettlinger G, Morton HB, Moffett E (1966) Tactile discrimination performance in the monkey. The effect of bilateral posterior parietal and lateral frontal ablations, and of callosal section. Cortex 2:5-29

Evarts EV (1966) Methods for recording activity of individual neurons in moving animals. In: Rushmer RF (ed) Methods in medical research. Year Book Medical Publishers, Chicago, I11, pp 241-250

Evarts EV (1968) A technique for recording activity of subcortical neurons in moving animals. Electroencephalogr Clin Neurophysiol 24:83-86

Eyssette M (1969) Le syndrom parieto-occipital bilateral. Tixier & fils, Lyon

Faugier-Grimaud S, Frenois C, Stein DG (1978) Effects of posterior parietal lesions on visually guided behavior in monkeys. Neuropsychologia 16:151-168

Ferrier D (1876) The functions of the brain. Smith Elder, London

Fischer B, Noth R (1981) Selection of visual targets activates prelunate cortical cells in trained rhesus monkeys. Exp Brain Res 41:431-433

Fleming JFR, Crosby EA (1955) The parietal lobe as an additional motor area. The motor effects of electrical stimulation and ablation of cortical area 5 and 7 in monkeys. J Comp Neurol 10:485-51

Foerster O (1931) The cerebral cortex in man. Lancet II:309-312

Foerster O (1936a) The motor cortex in man in the light of Hughlings Jackson's observations. Brain 59:135-159

Foerster O (1936b) Motorische Felder und Bahnen. In: Bumke O, Foerster O (eds) Handbuch der Neurologie, vol VI. Springer, Berlin Heidelberg New York, pp 1-352

Foerster O (1936c) Sensible corticale Felder. In: Bumke O, Foerster O (eds) Handbuch der Neurologie, vol VI. Springer, Berlin Heidelberg New York, pp 358-448

Fraiberg S (1977) Insights from the blind. Basic Books, New York

Fraiberg S, Freedman DA (1964) Studies in the ego development of the congenitally blind child. Psychoanal Study Child 19:113-169

French JD, Hernández-Peón R, Livingston RB (1955) Projections from cortex to cephalic brainstem (reticular formation) in monkey. J Neurophysiol 18:74-95

Friedman D, Jones EG, Burton H (1980) Representation pattern in the second somatic sensory area of the monkey cerebral cortex. J Comp Neurol 192:21-41

Friendlich AR (1973) Primate head restrainer using a nonsurgical technique. J Appl Physiol 35:934-935

Galaburda A, Sanides F (1980) Cytoarchitectonic organization of the human auditory cortex. J Comp Neurol 190:597-610

Gardner EP, Costanzo RM (1980) Neuronal mechanisms underlying direction sensitivity of somatosensory cortical neurons in awake monkeys. J Neurophysiol 4:1342-1354

Gerstmann J (1930) Zur Symptomatologie der Hirnläsionen im Übergangsgebiet der unteren Parietal- und mittleren Occipitalwindung. Nervenarzt 3:691-695

Godwin-Austen RB (1965) A case of visual disorientation. J Neurol Neurosurg Psychiatr 28:453-458

Goldberg ME (1980) Cortical mechanisms in the visual initiation of movement. Exp Brain Res 41:A32-33

Goldberg ME, Bushnell MC (1981) Behavioral enhancement of visual responses in monkey cerebral cortex. II. Modulation in frontal eye fields specifically related to saccades. J Neurophysiol 46:773-787

Goldberg ME, Robinson DL (1977) Visual responses of neurons in monkey inferior parietal lobule: the physiologic substrate of attention and neglect. Neurology 27:350

Goldberg ME, Robinson DL (1980) The significance of enhanced visual responses in posterior parietal cortex. Behav Brain Sci 3:503-505

Goodman LS, Gilman A (1970) The pharmacological basis of therapeutics, 4th edn. The Macmillan Co, London, p 97

Goodwin GM, McCloskey DI, Matthews PBC (1972) Proprioceptive illusions induced by muscle vibration: contribution by muscle spindles to perception? Science 175:1382-1384

Granit R (1972) Constant errors in the execution and appreciation of movement. Brain 95:649-660

Granit R (1977) The purposive brain. MIT Press, Cambridge, Mass

Graybiel AM (1970) Some thalamocortical projections of the pulvinar-posterior system of the thalamus in the cat. Brain Res 22:131-136

Gregory RL, Wallace JG (1963) Recovery from early blindness. A case study. Exp Psychol Monogr. Heffer & Sons, London

Grüsser O-J (1982) The multimodal structure of the extrapersonal space. In: Hein A, Jeannerod M (eds) Spatially oriented behavior. Springer, Berlin Heidelberg New York, in press

Haaxma R, Kuypers H (1974) Role of occipito-frontal cortico-cortical connections in visual guidance of relatively independent hand and finger movements in rhesus monkeys. Brain Res 71:361-366

Hartje W, Ettlinger G (1973) Reaching in light and dark after unilateral posterior parietal ablations in the monkey. Cortex 9:346-354

Hawrylyshyn PA, Rubin AM, Tasker RR, Organ LW, Fredrickson JM (1978) Vestibulothalamic projections in man – a sixth primary sensory pathway. J Neurophysiol 41:394-401

Headon MP, Powell TPS (1978) The effect of bilateral eye closure upon the lateral geniculate nucleus in infant monkey. Brain Res 143:147-154

Heath CJ, Jones EG (1970) Connexions of area 19 and the lateral suprasylvian area of the visual cortex of the cat. Brain Res 19:302-305

Hécaen H (1962) Clinical symptomatology in right and left hemispheric lesions. In: Mountcastle VB (ed) Interhemispheric relations and cerebral dominance. Johns Hopkins Press, Baltimore, pp 215-243

Hécaen H (1967) Brain mechanisms suggested by studies of parietal lobes. In: Brain mechanisms underlying speech mechanisms and language. Grune and Stratton, New York, pp 146-166

Hécaen H, Albert ML (1978) Human neuropsychology. John Wiley, New York

Hécaen H, Ajuriaguerra de J (1954) Balint's syndrome (psychic paralysis of visual fixation) and its minor forms. Brain 77:373-400

Hécaen H, Ajuriaguerra de J, Rouques L, David M, Dell MB (1950) Paralysie psychique du regard de Balint au cours de l'évolution d'une leucoencéphalite, type Balo. Rev Neurol 83:81-104

Hécaen H, Penfield W, Bertrand C, Malmo R (1956) The syndrome of apractognosia due to lesions of the minor cerebral hemisphere. Arch Neurol Psychiatr 75:400-434

Heilman KM (1979) Neglect and related disorders. In: Heilman KM, Valenstein E (eds) Clinical neuropsychology. Oxford University Press, Oxford

Heilman KM, Abell Van Den T (1980) Right hemisphere dominance for attention. The mechanisms underlying hemispheric asymmetries of inattention (neglect). Neurology 30:327-330

Heilman KM, Valenstein E (1972a) Auditory neglect in man. Arch Neurol 26:32-35

Heilman KM, Valenstein E (1972b) Frontal lobe neglect in man. Neurology 22:660-664

Heilman KM, Valenstein E (1979) Mechanisms underlying hemispatial neglect. Ann Neurol 5:166-170

Heilman KM, Watson RT (1977a) Mechanisms underlying the unilateral neglect syndrome. Adv Neurol 18:93-106

Heilman KM, Watson RT (1977b) The neglect syndrome – a unilateral defect of the orienting response. In: Harnad S, Doty RW, Goldstein L, Jaynes J, Krauthamer G (eds) Lateralization in the nervous system. Academic Press, London New York

Heilman KM, Pandya DN, Geschwind N (1970) Trimodal inattention following parietal lobe ablations. Trans Am Neurol Assoc 95:259-261

Heilman KM, Pandya DN, Karol EA, Geschwind N (1971) Auditory inattention. Arch Neurol 24:323-325

Heilman KM, Watson RT, Schulman HM (1974) A unilateral memory defect. J Neurol Neurosurg Psychiatr 37:790-793

Heilman KM, Schwartz HD, Watson RT (1978) Hypoarousal in patients with the neglect syndrome and emotional indifference. Neurology 28:229-232

Held R, Bauer JA (1967) Visually guided reaching in infant monkeys after restricted rearing. Science 155:718-720

Henn V, Young LR, Finley C (1974) Vestibular nucleus units in alert monkeys are also influenced by moving visual fields. Brain Res 71:144-149

Hernández-Peón R (1961) Reticular mechanisms of sensory control. In: Rosenblith WA (ed) Sensory communication. MIT Press, Cambridge, Mass, pp 497-520

Hernández-Peón R (1969) A neurophysiological and evolutionary model of attention. In: Evans CR, Mulholland TB (eds) Attention in neurophysiology. Butterworths, London, pp 417-426

Hernández-Peón R, Scherrer H, Jouvet M (1956) Modification of electrical activity in the cochlear nucleus during attention in unanesthetized cats. Science 123:331-332

Hier DB, Davis KR, Richardson EP, Mohr JP (1977) Hypertensive putaminal hemorrhage. Ann Neurol 1:152-159

Himwich HE, Callison DA (1972) The effects of alcohol on evoked potentials in various parts of the central nervous system of the cat. In: Kissin B, Begleiter H (eds) The biology of alcoholism, vol II. Plenum Press, New York, pp 67-84

Hof van-Duin van J (1976) Development of visuomotor behavior in normal and dark-reared cats. Brain Res 104:233-241

Holloway RL (1968) The evolution of the primate brain: some aspects of quantitative relations. Brain Res 7:121-172

Holloway RL (1974) The casts of fossil hominid brains. Sci Am 231:106-115

Holloway RL (1976) Paleoneurological evidence for language origins. Ann NY Acad Sci 280:330-348

Holloway RL (1978) The relevance of endocasts for studying primate brain evolution. In: Noback CR (ed) Sensory systems of primates. Plenum Press, New York, pp 181-200

Holmes G (1918) Disturbances of visual orientation. Br J Ophthalmol 2:449-468, 506-516

Holmes G (1919) Disturbances of visual space perception. Br Med J II:230-233

Holmes G, Horrax G (1919) Disturbances of spatial orientation and visual attention, with loss of stereoscopic vision. Arch Neurol Psychiatr 1:385-407

Holst von E, Mittelstaedt H (1950) Das Reafferenzprinzip (Wechselwirkungen zwischen Zentralnervensystem und Peripherie). Naturwissenschaften 37:464-476

Horn G (1963) The response of single units in the cortex of unrestrained cats to photic and somesthetic stimulation in the cat. J Physiol 165:80-81P

Hubel DH, Wiesel TN (1962) Receptive fields, binocular interaction and functional architecture in the cat's visual cortex. J Physiol 160:106-154

Hubel DH, Wiesel TN (1968) Receptive fields and functional architecture of monkey striate cortex. J Physiol 195:215-243

Hubel DH, Wiesel TN (1974) Sequence regularity and geometry of orientation columns in the monkey striate cortex. J Comp Neurol 158:267-294

Hubel DH, Wiesel TN (1977) Functional architecture of macaque monkey visual cortex. Proc R Soc London Ser B 198:1-59

Hyvärinen J (1973) Functional properties of cells in the parietal somatosensory and association cortices of the monkey. Acta Physiol Scand Suppl 396:36

Hyvärinen J (1980) Sensorimotor interaction in parietal association cortex. Behav Brain Sci 3:506-507

Hyvärinen J (1981a) Functional mechanisms of the parietal cortex. In: Grastyan E, Molnar P (eds) Advances in physiological sciences, plenary and symposia lectures of the 28th International Congress of Physiology, vol 16, Sensory functions. Pergamon Press and Academiai Kiado, Oxford and Budapest, pp 35-49

Hyvärinen J (1981b) Discussion on mechanisms of parietal cortex; how to study associative systems. In: Adam G, Meszaros I, Banyai EI (eds) Advances in physiological sciences, plenary and symposia lectures of the 28th International Congress of Physiology, vol 17, Brain and Behaviour. Pergamon Press and Academiai Kiado, Oxford and Budapest, pp 299-304

Hyvärinen J (1981c) Regional distribution of functions in parietal association area 7 of monkey. Brain Res 206:287-303

Hyvärinen J (1982) The posterior parietal lobe of the primate brain. Physiol Rev 62:1060-1129

Hyvärinen J, Hyvärinen L (1979) Blindness and modification of association cortex by early binocular deprivation in monkeys. Child 5:385-387

Hyvärinen J, Linnankoski I (1979) Breeding results in a colony of stumptail macaques (Macaca arctoides) used in physiological experiments. Primate Rep Nr 4:53-54

Hyvärinen J, Poranen A (1974) Function of the parietal associative area 7 as revealed from cellular discharges in alert monkeys. Brain 97:673-692

Hyvärinen J, Poranen A (1978a) Movement-sensitive and direction and orientation-selective cutaneous receptive fields in the hand area of the post-central gyrus in monkeys. J Physiol 283:523-537

Hyvärinen J, Poranen A (1978b) Receptive field integration and submodality convergence in the hand area of the post-central gyrus of the alert monkey. J Physiol 283:539-556

Hyvärinen J, Shelepin Yu (1979) Distribution of visual and somatic functions in the parietal associative area 7 of the monkey. Brain Res 169:561-564

Hyvärinen J, Sakata H, Talbot WH, Mountcastle VB (1968) Neuronal coding by cortical cells of the frequency of oscillating peripheral stimuli. Science 162:1130-1132

Hyvärinen J, Sakata H, Talbot WH, LaMotte RH, Mountcastle VB (1969) Periodic activity of postcentral neurons activated by sine wave mechanical stimuli of monkeys' hands compared with human frequency discrimination. Acta Physiol Scand Suppl 330:121

Hyvärinen J, Poranen A, Jokinen Y (1974) Central sensory activities between sensory input and motor output. In: Schmitt FO, Worden FG (eds) The neurosciences, 3rd study program. MIT Press, Cambridge, Mass, pp 205-214

Hyvärinen J, Poranen A, Jokinen Y, Näätänen R, Linnankoski (1975) Observations on unit activity in the primary somesthetic cortex of behaving monkeys. In; Kornhuber HH (ed) Somatosensory system. Thieme, Stuttgart, pp 241-249

Hyvärinen J, Hyvärinen L, Färkkilä M, Carlson S, Leinonen L (1978a) Modification of visual functions of the parietal lobe at early age in the monkey. Med Biol 56: 103-109

Hyvärinen J, Laakso M, Roine R, Leinonen L, Sippel H (1978b) Effect of ethanol on neuronal activity in the parietal association cortex of alert monkeys. Brain 101: 701-715

Hyvärinen J, Linnankoski I, Poranen A, Leinonen L, Altonen M (1978c) Use of monkeys as experimental animals: Report of a ten-year experience in a Nordic country. Ann Acad Sci Fenn A V Medica 172

Hyvärinen J, Laakso M, Roine R, Leinonen L (1979a) Comparison of effects of pentobarbital and alcohol on the cellular activity in the posterior parietal association cortex. Acta Physiol Scand 107:219-225

Hyvärinen J, Laakso M, Roine R, Leinonen L (1979b) Effects of phencyclidine, LSD and amphetamine on neuronal activity in the posterior parietal association cortex of the monkey. Neuropharmacol 18:237-242

Hyvärinen J, Poranen A, Jokinen Y (1980) Influence of attentive behavior on neuronal responses to vibration in primary somatosensory cortex of the monkey. J Neurophysiol 43:870-883

Hyvärinen J, Carlson S, Hyvärinen L (1981a) Early visual deprivation alters modality of neuronal responses in area 19 of monkey cortex. Neurosci Lett 26:239-243

Hyvärinen J, Hyvärinen L, Carlson S (1981b) Effects of binocular deprivation on parietal association cortex in young monkeys. Doc Ophthalmol Proc Ser 30:177-185

Hyvärinen J, Hyvärinen L, Linnankoski I (1981c) Modification of parietal association cortex and functional blindness after binocular deprivation in young monkeys. Exp Brain Res 42:1-8

Imig TJ, Ruggero MA, Kitzes LM, Javel E, Brugge JF (1977) Organization of auditory cortex in the owl monkey (Aortus trivirgatus). J Comp Neurol 171:11-128

Ingvar DH (1975) Patterns of brain activity revealed by measurements of regional cerebral blood flow. In: Ingvar DH, Lassen NA (eds) Brain work. Munksgaard, Copenhagen, pp 397-413

Ingvar DH, Rosen I, Eriksson M, Elmqvist D (1976) Activation patterns induced in the dominant hemisphere by skin stimulation. In: Zotterman Y (ed) Sensory functions of the skin in primates. Pergamon Press, Oxford, pp 549-559

Jampel RS (1960) Convergence, divergence, pupillary reactions and accommodation of the eyes from faradic stimulation of the macaque brain. J Comp Neurol 115:371-397

Jones EG, Powell TPS (1969) Connections of the somatic sensory cortex of the rhesus monkey. I. Ipsilateral cortical connections. Brain 92:477-502

Jones EG, Powell TPS (1970a) Connexions of the somatic sensory cortex of the rhesus monkeys. III. Thalamic connexions. Brain 93:37-53

Jones EG, Powell TPS (1970b) An anatomical study of converging sensory pathways within the cerebral cortex of the monkey. Brain 93:793-820

Jones EG, Powell TPS (1973) Anatomical organization of the somatosensory cortex. In: Iggo A (ed) Handbook of sensory physiology, Somatosensory system, vol II. Springer, Berlin Heidelberg New York, pp 578-620

Jung R (1974) Neuropsychologie und Neurophysiologie des Kontur- und Formsehens in Zeichnung und Malerei. In: Wieck HH (ed) Psychopathologie musischer Gestaltungen. Schattauer Verlag, Stuttgart, pp 29-88

Jung R, Kornhuber HH, Da Fonseca JS (1963) Multisensory convergence on cortical neurons. Neuronal effects of visual, acoustic and vestibular stimuli in the superior convolutions of the cat's cortex. Progn Brain Res 1:207-234

Kaas JH, Nelson RJ, Sur M, Lin C-S, Merzenich MM (1979) Multiple representations of the body within the primary somatosensory cortex of primates. Science 204:521-532

Kalant H (1975) Direct effects of ethanol on the nervous system. Fed Proc 34:1930-1941

Kase CS, Trancoso JF, Court JE, Tapia JF, Mohr JP (1977) Global spatial disorientation. Clinico-pathologic correlations. J Neurol Sci 34:267-278

Kawano K, Sasaki M (1981) Neurons in the posterior parietal association cortex of the monkey activated during optokinetic stimulation. Neurosci Lett 22:239-244

Kawano K, Sasaki M, Yamashita M (1980) Vestibular input to visual tracking neurons in the posterior parietal association cortex of the monkey. Neurosci Lett 17:55-60

Keller EL, Kamath BY (1975) Characteristics of head rotation and eye movement-related neurons in alert monkey vestibular nucleus. Brain Res 100:182-187

Kennard MA (1939) Alterations in response to visual stimuli following lesions of frontal lobe in monkeys. Arch Neurol Psychiatr 41:1153-116

Keys W, Robinson DL (1979) Eye movement-dependent enhancement of visual responses in the pulvinar nucleus of the monkey. Neurosci Abstr 5:791

Kling A, Orbach J (1963) The stump-tailed macaque: A promising laboratory primate. Science 139:45-46

Knudsen EI, Konishi M (1978) Center-surround organization of auditory receptive fields in the owl. Science 202:778-780

Kohonen T, Lehtiö P, Oja E (1981) Storage and processing of information in distributed associative memory systems. In: Hinton G, Anderson JA (eds) Parallel models of associative memory. Lawrence Erlbaum Associates, Hilsdale, New Jersey, pp 105-143

Künzle H (1978) Cortico-cortical efferents of primary motor and somatosensory regions of the cerebral cortex in Macaca fascicularis. Neuroscience 3:25-39

Kupfermann I, Weiss KR (1978) The command neuron concept. Behav Brain Sci 1: 3-39

Kuypers HGJM, Lawrence DG (1967) Cortical projections to the red nucleus and the brain stem in the rhesus monkey. Brain Res 4:151-188

LaMotte RH, Acuna C (1978) Defect in accuracy of reaching after removal of posterior parietal cortex in monkeys. Brain Res 139:309-326

Lanman J, Bizzi E, Allum J (1978) The coordination of eye and head movement during smooth pursuit. Brain Res 153:39-53

Latto R (1977) The effects of bilateral frontal eye-field, posterior parietal or superior collicular lesions on brightness thresholds in the rhesus monkey. Neuropsychologia 15:507-516

Latto R (1978) The effects of bilateral frontal eye-field, posterior parietal or superior collicular lesions on visual search in the rhesus monkey. Brain Res 146:35-50

Lawick-Goodall van J (1971) In the shadow in man. Collins, London

Leinonen L (1980) Functional properties of neurones in the posterior part of area 7 in awake monkey. Acta Physiol Scand 108:301-308

Leinonen L (1981) Functions of posterior parietal parietotemporal cortex in the monkey. Printed thesis, University of Helsinki, Dep Physiol

Leinonen LM, Hyvärinen J (1980) Parietal association cortex of the monkey as revealed by cellular recordings. In: Stelmach GE, Requin J (eds) Tutorials in motor behavior. North Holland, Amsterdam, pp 117-127

Leinonen L, Nyman G (1979) Functional properties of cells in anterolateral part of area 7, associative face area of awake monkey. Exp Brain Res 34:321-333

Leinonen L, Hyvärinen J, Nyman G, Linnankoski I (1979) Functional properties of neurons in lateral part of associative area 7 in awake monkey. Exp Brain Res 34: 299-320

Leinonen L, Hyvärinen J, Sovijärvi ARA (1980) Functional properties of neurons in the temporo-parietal association cortex of awake monkey. Exp Brain Res 39:203-215

Lemberger L, Rubin A (1976) Physiologic disposition of drugs of abuse. Spectrum Publications, New York, p 318

Leventhal AG, Hirsch HVB (1980) Receptive-field properties of different classes of neurons in visual cortex of normal and dark-reared cats. J Neurophysiol 43:1111-1132

Levine DN, Kaufman KJ, Mohr JP (1978) Inaccurate reaching associated with a superior parietal lobe tumor. Neurology 28:556-561

Liden CB, Lovejoy FH, Costello CE (1975) Phencyclidine. Nine cases of poisoning. JAMA 234:513-516

Lilly JC (1958) Correlations between neurophysiological activity in the cortex and short-term behavior in the monkey. In: Harlow HF, Woolsey CN (eds) Biological and biochemical bases of behavior. University of Wisconsin Press, pp 83-100

Lindsley DB (1960) Attention, consciousness, sleep and wakefulness. In: Field J, Magoun HW, Hall VE (eds) Handbook of physiology, section 1, neurophysiology, vol III. Am Physiol Soc, Washington DC, pp 1553-1593

Locke J (1690) An essay concerning human understanding. Wilburn R (ed) Everyman's library. Dent & Sons, London, pp 52-53

Luria AR (1959) Disorders of "simultaneous perception" in a case of bilateral occipito-parietal brain injury. Brain 82:437-449

Luria AR, Pravdina-Vinarskaya, Yarbuss AL (1963) Disorders of ocular movement in a case of simultanagnosia. Brain 86:219-228

Lynch JC (1980a) The role of parieto-occipital association cortex in oculomotor
control. Exp Brain Res 41:A32

Lynch JC (1980b). The functional organization of posterior parietal association cortex.
Behav Brain Sci 3:485-499

Lynch JC, McLaren JW (1979) Effects of lesions of parietooccipital association cortex
upon performance of oculomotor and attention tasks in monkeys. Neurosci Abstr
5:794

Lynch JC, McLaren JW (1982) Optokinetic nystagmus deficits following parieto-
occipital cortex lesions in monkeys. Exp Brain Res, in press

Lynch JC, Mountcastle VB, Talbot WH, Yin TCT (1977) Parietal lobe mechanisms
for directed visual attention. J Neurophysiol 40:362-389

Maciewicz RJ (1975) Thalamic afferents to area 17, 18 and 19 of cat cortex traced
with horseradish peroxidase. Brain Res 84:308-312

Marx JL (1973) Shortage of primates? Science 181:334

Matthews PBC (1982) Where does Sherrington's "muscular sense" originate? Muscles,
joints, corollary discharges? Annu Rev Neurosci 5:189-218

McCloskey DI, Torda TAG (1975) Corollary motor discharges and kinaesthesia. Brain
Res 100:467-470

McCulloch WS (1944) The functional organization of the cerebral cortex. Physiol
Rev 24:390-407

McDonnell PM (1979) Patterns of eye-hand coordination in the first year of life.
Canad J Psychol 33:253-267

McLaren JW, Lynch JC (1979) Quantitative studies of optokinetic nystagmus in
monkeys before and after lesions of parieto-occipital association cortex. Neurosci
Abstr 5:797

McLaren JW, Lynch JC (1980) Vertical optokinetic nystagmus deficit following
lesions of parieto-occipital association cortex in rhesus monkey. Neurosci Abstr
6:

Mendoza JE, Thomas RK (1975) Effects of posterior parietal and frontal neocortical
lesions in the squirrel monkey. J Comp Physiol Psychol 89:170-182

MacKay WA, Kwan MC, Murphy JT, Wong YC (1978) Responses to active and passive
wrist rotation in area 5 of awake monkeys. Neurosci Lett 10:235-239

Merzenich MM, Brugge JF (1973) Representation of the cochlear partition on the
superior temporal plane of the macaque monkey. Brain Res 50:275-296

Merzenich MM, Kaas JH, Sur M, Lin C-S (1978) Double representation of the body
surface within cytoarchitectonic areas 3b and 1 in "SI" in the owl monkey (Aotus
trivirgatus). J Comp Neurol 181:41-74

Mesulam M-M, Hoesen van GW, Pandya DN, Geschwind N (1977) Limbic and sensory
connections of the inferior parietal lobule (area PG) in the rhesus monkey: a study
with a new method for horseradish peroxidase histochemistry. Brain Res 136:
393-414

Michel F, Eyssette M (1972) L'ataxie optique et l'ataxie du regard dans les lésions
bilaterales de la jonction pariéto-occipitale. Rev Oto-Neuro-Ophthalmol (Paris)
2:177-186

Michel F, Jeannerod M, Devic M (1965) Trouble de l'orientation visuelle dans les trois
dimensions de l'espace. Cortex 1:441-466

Milner AD, Ockleford EM, Dewar W (1977) Visuo-spatial performance following
posterior parietal and lateral frontal lesions in stumptail macaques. Cortex 13:
350-360

Moffett AM, Ettlinger G (1970) Tactile discrimination performance in the monkey:
The effects of unilateral posterior parietal ablations. Cortex 6:47-67

Moffett AM, Ettlinger G, Morton HB, Piercy MF (1967) Tactile discrimination performance in the monkey: The effect of ablation of various subdivisions of posterior
parietal cortex. Cortex 3:59-96

Moll L, Kuypers HGJM (1977) Premotor cortical ablations in monkeys: Contralateral
changes in visually guided reaching behavior. Science 198:317-319

Moskowitz H (1973) Proceedings: Psychological tests and drugs. Pharmakopsychiatrie Neuropsychopharmakologie 6:114-126

Motter BC, Mountcastle VB (1981) The functional properties of the light sensitive neurons of the posterior parietal cortex studied in waking monkeys: foveal sparing and opponent vector orientation. J Neurosci 1:3-26

Mountcastle VB (1957) Modality and topographic properties of single neurons of cat's somatic sensory cortex. J Neurophysiol 20:408-434

Mountcastle VB (1978) An organizing principle for cerebral function: The unit module and the distributed system. In: Edelman GM, Mountcastle VB (eds) The mindful brain. MIT Press, Cambridge, Mass, pp 7-50

Mountcastle VB (1981) Functional properties of the light sensitive neurons of the posterior parietal cortex and their regulation by state controls: influence on excitability of interested fixation and the angle of gaze. In: Pompeiano O, Ajmone Marsan C (eds) Brain mechanisms of perceptual awareness and purposeful behavior. JBRO Monograph Series, Vol. 8. Raven Press, New York, pp 67-69

Mountcastle VB, Talbot WH, Sakata H, Hyvärinen J (1969) Cortical neuronal mechanisms in flutter-vibration studied in nonanesthetized monkeys. Neuronal periodicity and frequency discrimination. J Neurophysiol 32:452-484

Mountcastle VB, Lynch JC, Georgopoulos A, Sakata H, Acuna C (1975) Posterior parietal association cortex of the monkey: command functions for operations within extrapersonal space. J Neurophysiol 38:871-908

Mountcastle VB, Motter BC, Anderson RA (1980) Some further observations on the functional properties of neurons in the parietal lobe of the waking monkey. Behav Brain Sci 3:520-523

Mountcastle VB, Andersen RA, Motter BC (1981) The influence of attentive fixation upon the excitability of the light sensitive neurons of the posterior parietal cortex. J Neurosci 1:1218-1235

Munk H (1881) Über die Funktionen der Grosshirnrinde. Hirschwald, Berlin

Murata K, Cramer H, Bach-y-Rita P (1965) Neuronal convergence of noxious, acoustic, and visual stimuli in the visual cortex of the cat. J Neurophysiol 28:1223-1239

Noorden von GK, Crawford MLJ (1978) Morphological and physiological changes in the monkey visual system after short-term lid suture. Invest Ophthalmol Vis Sci 17:762-768

Orem J, Schlag-Rey M, Schlag J (1973) Unilateral visual neglect and thalamic intralaminar lesions in the cat. Exp Neurol 40:784-797

Pandya DN, Kuypers HGJM (1969) Cortico-cortical connections in the rhesus monkey. Brain Res 13:13-36

Pandya DN, Sanides F (1973) Architectonic parcellation of the temporal operculum in rhesus monkey and its projection pattern. Z Entwicklungsgesch 139:127-161

Pandya DN, Selzer B (1982) Intrinsic connections and architectonics of posterior parietal cortex in the rhesus monkey. J Comp Neurol 204:196-210

Passingham RE, Ettlinger G (1974) A comparison of cortical functions in man and the other primates. Int Rev Neurobiol 16:233-299

Paterson A, Zangwill OL (1944) Disorders of visual space perception associated with lesions of the right cerebral hemisphere. Brain 67:331-358

Paul RL, Merzenich M, Goodman H (1972) Representation of slowly and rapidly adapting cutaneous mechanoreceptors of the hand in Brodmann's areas 3 and 1 of Macaca mulatta. Brain Res 36:229-249

Pearson RCA, Brodal P, Powell TPS (1978) The projection of the thalamus upon the parietal lobe in the monkey. Brain Res 144:143-148

Peele TL (1942) Cytoarchitecture of individual parietal areas in the monkey (Macaca mulatta) and the distribution of the efferent fibers. J Comp Neurol 77:693-738

Peele TL (1944) Acute and chronic parietal lobe ablations in monkeys. J Neurophysiol 7:269-286

Penfield W, Rasmussen T (1950) The cerebral cortex of man. Macmillan, New York

Perrin RG, Hockman CH, Kalant H, Livingston KE (1974) Acute effects of ethanol on spontaneous and auditory evoked electrical activity in cat brain. Electroencephalogr Clin Neurophysiol 36:19-31

Perry J, Nickel VL (1959) Total cervical spine fusion for neck paralysis. J Bone Joint Surg 41-A:37-59

Petras JM (1971) Connections of the parietal lobe. J Psychiatr Res 8:189-201

Petrides M, Iversen SD (1979) Restricted posterior parietal lesions in the rhesus monkey and performance on visuospatial tasks. Brain Res 161:63-77

Pettigrew JD (1974) The effect of visual experience on the development of stimulus specificity by kitten cortical neurones. J Physiol 237:49-74

Phillips CG, Powell TPS, Wiesendanger M (1971) Projection from low-threshold muscle afferents of hand and forearm to area 3a of baboons cortex. J Physiol 217:419-446

Pohl W (1973) Dissociation of spatial discrimination deficits following frontal and parietal lesions in monkeys. J Comp Physiol Psychol 82:227-239

Poranen A, Hyvärinen J (1982) Effects of attention on multiunit responses to vibration in the somatosensory regions of the monkey's brain. Electroencephalogr Clin Neurophysiol, 53:525-537

Powell TPS, Mountcastle VB (1959a) The cytoarchitecture of the postcentral gyrus of the monkey Macaca mulatta. Bull Johns Hopkins Hosp 105:108-131

Powell TPS, Mountcastle VB (1959b) Some aspects of the functional organization of the cortex of the postcentral gyrus of the monkey: a correlation of findings obtained in a single unit analysis with cytoarchitecture. Bull Johns Hopkins Hosp 105:133-162

Pribram H, Barry J (1956) Further behavioral analysis of parieto-temporo-preoccipital cortex. J Neurophysiol 19:99-106

Prolo DJ, Runnels JB, Jameson RM (1973) The injured cervical spine. Immediate and long-term immobilization with the halo. J Am Med Assoc 224:591-594

Randolph M, Semmes J (1974) Behavioral consequences of selective subtotal ablations in the postcentral gyrus of Macaca mulatta. Brain Res 70:55-70

Ratcliff G, Davies-Jones GAB (1972) Defective visual localization in focal brain wounds. Brain 95:49-60

Ratcliff G, Ridley RM, Ettlinger G (1977) Spatial disorientation in the monkey. Cortex 13:62-65

Reeves AG, Hagaman WD (1971) Behavioral and EEG asymmetry following unilateral lesions of the forebrain and midbrain in cats. Electroencephalogr Clin Neurophysiol 30:83-86

Regal DM, Boothe R, Teller DY, Sackett GP (1976) Visual acuity and visual responsiveness in dark-reared monkeys (Macaca nemestrina). Vision Res 16:523-530

Riddoch G (1935) Visual disorientation in homonymous half-fields. Brain 58:376-382

Riesen AH (1958) Plasticity of behavior: Psychological aspects. In: Harlow H, Woolsey CN (eds) Biological and biochemical bases of behavior. University of Wisconsin Press, Madison, Wis, pp 425-450

Riesen AH (1961a) Studying perceptual development using the technique of sensory deprivation. J Nerv Ment Dis 132:21-25

Riesen AH (1961b) Stimulation as requirement for growth and function in behavioral development. In: Fiske DW, Maddi SR (eds) Functions of varied experience. Dorsey Press, Homewood, Ill, pp 57-105

Riesen AH (1966) Sensory deprivation. In: Stellar E, Sprague JM (eds) Progress in physiological psychology, vol I. Academic Press, London New York, pp 117-147

Riesen AH, Aarons L (1959) Visual movement and intensity discrimination in cats after early deprivation of pattern vision. J Comp Physiol Psychol 52:142-149

Riesen AH, Ramsey RL, Wilson PD (1964) Development of visual acuity in rhesus monkeys deprived of patterned light during early infancy. Psychon Sci 1:33-34

Rizzolatti G, Scandolara C, Matelli M, Gentilucci M (1981) Afferent properties of periarcuate neurons in macaque monkeys. II. Visual responses. Behav Brain Res 2: 147-163

Robinson CJ, Burton H (1980a) Somatotopographic organization in the second somatosensory area of M. fascicularis. J Comp Neurol 192:43-67

Robinson CJ, Burton H (1980b) The organization of somatosensory receptive fields in cortical areas 7b, retroinsular, postauditory and granular insular of M. fascicularis. J Comp Neurol 192:69-92

Robinson CJ, Burton H (1980c) Somatic submodality distribution within the second somatosensory (SII), 7b, retroinsular, postauditory, and granular insular cortical areas of M. fascicularis. J Comp Neurol 192:93-108

Robinson DL, Keys W (1980) Visuo-motor properties of neurons in superior colliculus and pulvinar of the monkey. Proc Int Un Physiol Sci 14, 28 Int Congr Budapest, 226

Robinson DL, Goldberg ME, Stanton GB (1978) Parietal association cortex in the primate: sensory mechanisms and behavioral modulations. J Neurophysiol 41: 910-932

Roland PE, Skinhoj E, Lassen NA, Larsen B (1980) Different cortical areas in man in organization of voluntary movements in extrapersonal space. J Neurophysiol 43:137-150

Rolls ET, Perret D, Thorpe SJ, Puerto A, Roper-Hall A, Maddison S (1979) Responses of neurons in area 7 of the parietal cortex to objects of different significance. Brain Res 169:194-198

Rolls ET, Perret D, Thorpe SJ (1980) The influence of motivation on the responses of neurons in the posterior parietal association cortex. Behav Brain Sci 3:514-515

Rondot P, Recondo de J (1974) Ataxie optique: Trouble de la coordination visuo-motrice. Brain Res 71:367-375

Rosenquist AC, Edwards SB, Palmer LA (1974) An autoradiographic study of the projections of the dorsal lateral geniculate nucleus and the posterior nucleus in the cat. Brain Res 80:71-93

Roth M (1949) Disorders of body image caused by lesions of the right parietal lobe. Brain 72:89-111

Rovamo J, Hyvärinen J (1976) A noiseless model of associative memory based on the cortical structure and on the effect of the state of consciousness. Exp Brain Res Suppl 1:484-489

Ruch TC, Fulton JF, German WJ (1938) Sensory discrimination in monkey, chimpanzee and man after lesions of the parietal lobe. Arch Neurol Psychiatr 39:919-938

Sakata H (1975) Somatic sensory responses of neurons in the parietal association area (area 5) in monkeys. In: Kornhuber HH (ed) The somatosensory system. Thieme, Stuttgart, pp 250-261

Sakata H, Kawano K (1982) Role of parietal association cortex in space perception and motor guidance. Annu Rev Neurosci, in print

Sakata H, Takaoka A, Kawarasaki A, Shibutani H (1973) Somatosensory properties of neurons in superior parietal cortex (area 5) of the rhesus monkey. Brain Res 64:85-102

Sakata H, Shibutani H, Kawano K (1978) Parietal neurons with dual sensitivity to real and induced movements of visual target. Neurosci Lett 9:165-169

Sakata H, Shibutani H, Kawano K (1980) Spatial properties of visual fixation neurons in posterior parietal association cortex of the monkey. J Neurophysiol 43:1654-1672

Samuels I, Butters N, Goodglass H (1971) Visual memory deficits following cortical and limbic lesions: effect of field of presentation. Physiol Behav 6:447-452

Saraux H, Esteve P, Graveleau D, Goupil H (1962) Syndrome de Balint et apraxie oculo-motrice. Ann Ocul 195:456-472

Schilder P, Stengel E (1928) Schmerzasymbolie. Z Ges Neurol Psychiatr 113:143-158

Schilder P, Stengel E (1931) Asymbolia for pain. Arch Neurol Psychiatr 25:598-600

Schneider GE (1969) Two visual systems. Science 163:895-902

Schwarz DWF, Fredrickson JM (1971a) Tactile direction sensitivity of area 2 oral neurons in the rhesus monkey cortex. Brain Res 27:397-401

Schwarz DWF, Fredrickson JM (1971b) Rhesus monkey vestibular cortex: A bimodal primary projection field. Science 172:280-281

Segal A, Stone FH (1961) The six-year-old who began to see: emotional sequelae of operation for congenital bilateral cataracts. Psychoanal Study Child 16:481-509

Selzer B, Pandya DN (1978) Afferent cortical connections and architectonics in the superior temporal sulcus and surrounding cortex in the rhesus monkey. Brain Res 149:1-24

Selzer B, Pandya DN (1980) Converging visual and somatic sensory cortical input to the interparietal sulcus of the rhesus monkey. Brain Res 192:339-351

Semmes J, Turner B (1977) Effects of cortical lesions on somatosensory tasks. J Invest Dermatol 69:181-189

Semmes J, Weinstein S, Ghent L, Teuber H-L (1963) Correlates of impaired orientation in personal and extrapersonal space. Brain 86:747-772

Semmes Blum J, Chow KL, Pribram KH (1950) A behavioral analysis of the organization of the parieto-temporo-preoccipital cortex. J Comp Neurol 93:53-100

Senden von M (1932) Raum- und Gestaltauffassung bei operierten Blind-geborenen vor und nach der Operation. Barth, Leipzig. English translation: Space and sight. Free Press, Glencoe, Ill, 1960

Shibutani H, Sakata H, Kawano K (1982) Functional properties of visual tracking neurons in the posterior parietal association cortex of the monkey. J Neurophysiol, in print

Smith JL (1963) Optokinetic nystagmus. Thomas, Springfield, pp 55-67

Smith JL, Cogan DG (1959) Optokinetic nystagmus: a test for parietal lobe lesions. Am J Ophthalmol 48:187-193

Sokolov Ye N (1963) Perception and the conditioned reflex. Pergamon Press, Oxford

Sperry RW (1950) Neural basis of the spontaneous optokinetic response produced by visual inversion. J Comp physiol Psychol 43:482-489

Stanton GB, Cruce WLR, Goldberg ME, Robinson DL (1977) Some ipsilateral projections to area PF and PG of the inferior parietal lobule in monkeys. Neurosci Lett 6:243-250

Stein J (1976) The effect of cooling parietal lobe areas 5 and 7 upon voluntary movement in awake rhesus monkeys. J Physiol 258:62-63P

Stein J (1978) The effect of parietal lobe cooling on manipulative behaviour in the conscious monkey. In: Gordon G (ed) Active touch. Pergamon Press, Oxford, pp 79-90

Strampelli B, Valvo A, Scholler H (1969) Probleme des visuellen Lernprozesses nach langjähriger Blindheit. Horus. Marburger Beitr Blind-Sehen 1:1-6

Sugishita M, Ettlinger G, Ridley RM (1978) Disturbance of cage-finding in the monkey. Cortex 14:431-438

Sundqvist A (1979) Saccadic reaction-time in parietal-lobe dysfunction. Lancet I:870

Sur M, Merzenich MM, Kaas JH (1980) Magnification, receptive field area, and "hypercolumn" size in areas 3b and 1 of somatosensory cortex in owl monkeys. J Neurophysiol 44:295-311

Sur M, Wall JT Kaas JH (1981) Modular segregation of functional cell classes within the postcentral somatosensory cortex of monkeys. Science 212:1059-1061

Suzuki H, Azuma M (1977) Prefrontal neuronal activity during gazing at a light spot in the monkey. Brain Res 126:497-508

Szentagothai J (1975) The 'module-concept' in cerebral cortex architecture. Brain Res 95:475-496

Talbot WH, Darian-Smith I, Kornhuber HH, Mountcastle VB (1968) The sense of flutter-vibration: comparison of the human capacity with response patterns of mechanoreceptive afferents from the monkey hand. J Neurophysiol 31:301-334

Teuber HL (1963) Space perception and its disturbances after brain injury in man. Neuropsychologia 1:47-57

Thompson RF, Johnson RH, Hoopes JJ (1963) Organization of auditory, somatic sensory and visual projection to association fields of cerebral cortex in cat. J Neurophysiol 26:343-364

Timney B, Mitchell DE, Giffin F (1978) The development of vision in cats after extended periods of dark-rearing. Exp Brain Res 31:547-560

Trevarthen CB (1968) Two mechanisms of vision in primates. Psychol Forsch 31:299-337

Trojanowski JQ, Jacobson S (1977) Areas and laminar distribution of some pulvinar cortical efferents in rhesus monkey. J Comp Neurol 169:371-392

Tyler HR (1968) Abnormalities of perception with defective eye movements (Balint's syndrome). Cortex 4:154-171

Ungerleider LG, Brody BA (1977) Extrapersonal spatial orientation: The role of the posterior parietal, anterior frontal, and infero-temporal cortex. Exp Neurol 56:265-280

Valvo A (1971) Sight restoration after long-term blindness. Am Found for the Blind, New York

Vogt BA, Pandya DN (1978) Cortico-cortical connections of somatic sensory cortex (areas 3, 1 and 2) in the rhesus monkey. J Comp Neurol 177:179-192

Vogt C, Vogt O (1919) Allgemeine Ergebnisse unserer Hirnforschung. J Psychol Neurol 25:279-462

Vogt C, Vogt O (1926) Die vergleichend-architektonische und die vergleichend-reiz-physiologische Felderung der Grosshirnrinde unter besonderer Berücksichtigung der menschlichen. Naturwissenschaften 14:1190-1194

Waespe W, Henn V (1977) Neuronal activity in the vestibular nuclei of the alert monkey during vestibular and optokinetic stimulation. Exp Brain Res 27:523-538

Wagman IH (1964) Eye movements induced by electric stimulation of cerebrum in monkeys and their relationship to bodily movements. In: Bender MB (ed) Oculomotor system. Hoeber, New York, pp 18-39

Warren DH (1977) Blindness and early childhood development. Am Found for the Blind, New York

Watson RT, Heilman KM (1979) Thalamic neglect. Neurology 29:690-694

Watson RT, Heilman KM, Cauthen JC, King FA (1973) Neglect after cingulectomy. Neurology 23:1003-1007

Watson RT, Heilman KM, Miller BD, King FA (1974) Neglect after mesencephalic reticular formation lesions. Neurology 24:294-298

Watson RT, Miller B, Heilman KM (1978) Nonsensory neglect. Ann Neurol 3:505-508

Weinstein EA, Kahn RL, Slote W (1955) Withdrawal, inattention and pain asymbolia. Arch Neurol Psychiatr 74:235-248

Welch K, Stuteville P (1958) Experimental production of unilateral neglect in monkeys. Brain 81:341-347

Werner G, Whitsel BL (1968) Topology of the body representation in somatosensory area I of primates. J Neurophysiol 31:856-869

Whitfield IC (1971) Mechanisms of sound source localization. Nature (London) 233:95-97

Whitsel BL, Petrucelli LM, Werner G (1969) Symmetry and connectivity in the map of the body surface in somatosensory area II of primates. J Neurophysiol 32:170-183

Whitsel BL, Dreyer DA, Roppolo JR (1971) Determinants of body representation in postcentral gyrus of macaques. J Neurophysiol 34:1018-1034

Whitsel BL, Roppolo JR, Werner G (1972) Cortical information processing of stimulus motion on primate skin. J Neurophysiol 35:691-717

Whitty CWM, Newcombe F (1965) Disabilities associated with lesions in the posterior parietal region of the non-dominant hemisphere. Neuropsychologia 3:175-185

Wiesel TN, Hubel DH (1963) Single.cell responses in striate cortex of kittens deprived of vision in one eye. J Neurophysiol 26:1003-1017

Wiesel TN, Hubel DH (1965) Comparison of the effects of unilateral and bilateral eye closure on cortical unit responses in kittens. J Neurophyisol 28:1029-1040

Wilson M (1957) Effects of circumscribed cortical lesions upon somesthetic and visual discrimination in the monkey. J Comp Physiol Psychol 50:630-635

Wilson M, Stamm JS, Pribram KH (1960) Deficits in roughness discrimination after posterior parietal lesions in monkeys. J Comp Physiol Psychol 53:535-539

Wolpert I (1924) Die Simultanagnosie − Störung der Gesamtauffassung. Z Ges Neurol Psychiatr 93:397-415

Woolsey CN, Fairman D (1946) Contralateral, ipsilateral and bilateral representation of cutaneous receptors in somatic areas I and II of the cerebral cortex of pig, sheep and other mammals. Surgery 19:684-702

Woolsey CN, Marshall WH, Bard P (1942) Representation of cutaneous tactile sensibility in the cerebral cortex of the monkey as indicated by evoked potentials. Bull Johns Hopkins Hosp 70:399-441

Yin TCT, Mountcastle VB (1977) Visual input to the visuomotor mechanisms of the monkey's parietal lobe. Science 197:1381-1383

Young LR, Dichgans J, Murphy R, Brandt Th (1973) Interaction of optokinetic and vestibular stimuli in motion perception. Acta Oto-Laryngol 76:24-31

Zeki SM (1978) Functional specialization in the visual cortex of the rhesus monkey. Nature (London) 274:423-428

Subject Index

Advanced Views in Primate Biology

Main Lectures of the VIIIth Congress of the International Primatological Society, Florence, July 7–12, 1980

Editors: **A.B.Chiarelli, R.S.Corruccini**

1982. 35 figures, 36 tables. XX, 266 pages (Proceedings in Life Sciences)
ISBN 3-540-11092-4

From the contents: Part A – Main Lectures: Recent Advances in Molecular Evolution and Immunogenetic Evolution of Primates. – Evolution of Human Skin. – Evolution of Language and Intelligence in Hominids. – Primatology and Sociobiology. – Sexual Behavior in Aging Male Rhesus Monkeys. – Simiantype Blood Groups of Hamadryas Baboons. – The Role of a Kenyan Primate Center in Conservation. – Further Declines in Rhesus Populations of India. – Taiwan Macaques: Ecology and Conservation Needs. – Prospects for a Self-sustaining Captive Chimpanzee. Breeding Program. – Part B – Symposium Reports: Miocene Hominoids and New Interpretations of Ape and Human Ancestry. – Infanticide in Langue Monkeys. – Recent Advances in the Study of Tool-use Non-human Primates. – Summary of the Satellite Symposium on Primate Communication. – Primate Locomotor Systems. – "Methods and Concepts in Primate Brain Evolution". – The Effects of Drugs and Hormones on Social Behavior in Non-human Primates. – Chromosome Banding and Primate Phylogeny. – The Present and Future Status of Comparative Psychology.

Axoplasmic Transport

Editor: **D.G.Weiss**

1982. 181 figures. Approx. 400 pages (Proceedings in Life Sciences)
ISBN 3-540-11662-1

Contents: Non-Neuronal Intracellular Motility. – Molecular and Structural Components of the Axon: Boundary Conditions for Transport. – General Characterization of Axoplasmic Transport: Materials and Properties. – Transport Mechanism Prerequisites: Experimental Approaches. – Transport Mechanisms: Theoretical Approaches and Models. – Experimental Techniques to Study Axoplasmic Transport.

V.Braitenberg

On the Texture of Brains

An Introduction to Neuroanatomy for the Cybernetically Minded

Translated from the German by E.H.Braitenberg and by the author
Heidelberg Science Library
1977. 37 figures. IX, 127 pages
ISBN 3-540-08391-X

Contents: Neuroanatomy, Psychology, and Animism. – Physics and Antiphysics. – Information. – What Brains Are Made Of? – How Accurately Are Brains Designed? – Neuroanatomical Invariants: Analysis of the Cerebellar Cortex. – The Automatic Pilot of the Fly. – The Common Sensorium: An Essay on the Cerebral Cortex.

Springer-Verlag Berlin Heidelberg New York